THE VANDEMONIAN

Wall Street and Silicon Valley Collide

ALLAN CHARLES BRANCH

First published 2025

Big Sky Publishing Pty Ltd
PO Box 303, Newport, NSW 2106, Australia
Phone: 1300 364 611
Email: info@bigskypublishing.com.au
Web: www.bigskypublishing.com.au

Cover design and typesetting: Think Productions

A catalogue record for this book is available from the National Library of Australia

Title: The Vandemonian: Wall Street and Silicon Valley Collide
ISBN: 978-1-923300-22-4

THE VANDEMONIAN

Wall Street and Silicon Valley Collide

www.bigskypublishing.com.au

ALLAN CHARLES BRANCH

CONTENTS

'One crowded hour being worth more than a lifetime of the commonplace or the mundane.'

Moss Webster, in *Life or Death* by Michael Robotham

INTRODUCTION

A LIFE OF SOARING

Life is unsettling with few soft landings. But to have soft landings you first have to soar.

Many lives are spent putting one step in front of the other, unthinking, getting from here to there. My life has been in three distinct stages, each compartmentalised, but not completely disconnected. There are doors funnelling between each compartment because each has been a collision with the next. More tiered than stages, no forks in the road, no points of decision. It's like going from room to room at the Versailles Palace, whose designers forgot to add corridors. Going directly from the Antechamber through the King's Chamber on through to the Council Chamber.

At first was a junior's internship of doing anything and everything in my home state of Tasmania, Australia; a training ground of working in government, business, large and small employers, employing myself, included some university education. Discovering who I wasn't.

Followed by almost 20 years of discovering who I wanted to be. Inventing, prototyping, manufacturing, selling and promoting the smartest mobile robots worldwide, to the extent that at the time I left the industry, it is probable that I had designed and built and sold more autonomous robots than anyone on the planet. My collision with Silicon Valley.

Then another 20 years of turning around stressed companies, of any kind with any problem, anywhere. And not only stressed but seemingly irredeemably failed; all saved such was the effectiveness of the problem-solving techniques I uncovered and developed, finally discovering who I was. A collision with the world's financial, investment and banking institutions.

Looking back, I cannot reconcile that unextraordinary boy from a modest working-class family in Tasmania with the one who came to live and work in some 20 countries on different continents and circled the globe a hundred times or more. It is such an implausible, non-linear story: one difficult to document chronologically.

So, my memoir is a series of vignettes, reminiscences, thoughts, observations, themes of personal development and personal failings. Soaring for sure – living large – but colliding definitely.

In the beginning, what I wasn't, was settled. In the middle, what I wanted was to determine my potential, and in the end, what I was, was satisfied.

'My story … is about progress,
not always in the direction intended.'

CHAPTER 1

NIPALUNA

Australia was discovered by a skilled and brave but unknown explorer about 60,000 years ago. Took his family along for the ride. Nobody quite knows where from; Borneo, New Guinea, Africa are all candidates, but there is no genetic link to any of those populations. Over time his generations spread across the wide flat continent as far south as Tasmania. Back then it was one land mass, oceans lower, land bridges not flooded, so walking was okay; around the coastline and along river courses was sensible because the inhospitable interior was already a vast, treeless desert, the inland sea and lushness having disappeared a couple of million years earlier.

Along the way the family and their families left signs of their existence and activities and lives, so there is no doubt. There is language, culture, art, middens, tools, weapons, kitchen utensils, maybe even clay pottery, trade, and as uncomfortable as it is to say, extinction of some of the wildlife which had never been threatened by humans before. We like to think of Europeans as the cause of wildlife extinction, plants as well as animals, but it seems to be a universal consequence of the human talent at hunting and killing. Not tigers or bears or large predators in Australia, but giant animals nevertheless; megafauna, oversize wombats and kangaroos the size of cattle, all disappeared, presumably to the dinner plates. I don't think plates have been found. That meant having the knowledge of and use of fire, not just to cook up a feast, but to manipulate the environment, something nature was already good at in dry, hot, wildfire-dominated Australia, and which the Aborigines adopted. In other words, they observed how nature sustained itself, and then duplicated those methods. Europeans ignored all of that and brought European plants, animals and farming methods to this resistant, still defiant land.

He was no more than medium height, black skin, robust and a skilled hunter-fisher. Instantly, nothing was recognisable to him in his new land. Just a few miles from his stepping off point, probably Indonesia, across that invisible Wallace Line with its impenetrably deep-water chasm that caused millions of years of separate evolution, the Eurasian ecology became Australian ecology. They could have been different planets. Along the way, slowly, he learned of the new plants: eucalyptus, new animals: marsupials, no massive predators; new seasons and climate:

becoming dry continental and therefore even more unrecognisable as he moved from the coast to the blazing days and freezing nights of never-ending desert, then temperate and cool as he progressed further south. He learned to hunt the different animals, to small-farm the plants, control the forests, protect himself and his family. His past became vague and twisted like a dream, evolving into an origin story he called the Dreamtime, the world's first meme.

When we learned about all of this at school, one thing I remember distinctly was the teacher telling us about the low Aboriginal population numbers and how the Aborigines had low fertility rates. I cannot remember how old I was, and fertility was just a comprehendible concept at whatever age it was. The Eurocentric ignorance in that statement by a white, biased, English-ish role model is apparent now, but it was accepted doctrine then. Unquestioned. The possibility that a smaller population was stable and sustainable and preferable to the wanton exponential growth that threatens the world today was inconceivable when I was a kid, when parts of my own state had not even been explored or mapped, remaining uncharted, terra incognito, and the world was still infinite. I mean, they existed comfortably for 60,000 years, content with local resources, implementing sustainability before anyone ever heard the word. With plenty of land and no pressure there is no evidence that they were driven to colonise anyone else's land.

Other waves of peoples seemed to have arrived, through South-East Asia most likely, more genes and ideas, complicating any understanding of the convoluted history; for instance, bringing dingoes with them one time, about 4000 years ago.

In Tasmania, the indigenous peoples became trapped because of global warming. Not because they mined fossil fuels and raised herds of grazing animals, or burned down forests, although they certainly did that, but because the last ice age ended and the oceans rose. So Tasmanian Aborigines have lived here alone and isolated for about 14,000 years and are still here.

The dingoes came too late to make it across the land bridge to Tasmania, else they would surely have eradicated the thylacine, Australia's largest carnivore, the Tasmanian tiger, like they did on the mainland. That was left to the Europeans to do within a century or so of settlement, after the animal had survived quite adequately for 160 million years. The last thylacine died in captivity in a primitive zoo in 1936 in Hobart, the capital city of Tasmania, a city named after a British warlord. The reconstructed palawa kani Aboriginal word for Hobart, where I spent my formative years, is nipaluna.

I have one thirty-second Tasmanian Aboriginal blood. By someone's rules that is enough to qualify as an Aborigine, and I am proud of it, but I only found out later in life, so it does not morally count.

Humans were not the only living things trapped here. So, too, were animals like the thylacine: Tasmanian Devil; the local emu subspecies also hunted to extinction by Europeans; anaspides, a shrimp so ancient they had not bent their tails yet; the equally ancient cave spider aptly named *Hickmania troglodytes*; the fagus beech, the only Australian deciduous plant; the Tasmanian blue gum (*Eucalyptus globulus*), the world's tallest flowering trees, possibly the tallest of any trees; the useful Huon pine. Then there are the really odd ones, like cave-dwelling glow worms that are not worms at all but insects.

Australia is famous for its mammals, defined by having hair on their bodies, and the females producing milk for their babies. All mammals are divided into three groups. Some, like us, are lucky enough to give birth to well-developed babies. Some are able to survive without assistance from birth, like those deer whose newborns can walk upright almost immediately – these are named placentals. Their babies gestate in a perfect womb environment nourished by a placenta, which is a filter to extract sustenance from the mother's blood supply and pass it on to the foetus.

A little more primitive, but still with live born babies, are the famous marsupials of Australia. Baby joeys so underdeveloped that for a long time they had not been seen, researchers mistakenly believing that the neonates grew off the tip of the mother's teat. They start gestation in a primitive womb, crawling out when about the size of a jellybean, some as small as a grain of rice, finding the nipple in the pouch by sniffing the smell of milk, then continuing to grow for the rest of the time externally and seemingly attached.

Even more primitive are the mammals that actually lay eggs. Like birds, and reptiles and fish and most other things. They do still supply mother's milk to the babies, which are called puggles, just through skin pores. It is a toss-up; are they mammals? They have hair instead of feathers or scales, and that milk from the skin, so are definitely mammals.

The exits on the bodies of this last group of animals are combined: urine, faeces and eggs all emerge from the same opening. One hole, or if you want to be clever, a monotreme. The opening is called a cloaca and the Australian slang word 'clacker', used in the expression, 'up your clacker', is picturesquely derivative. There are only two monotreme types left, both in Tasmania: the echidna and the platypus. We used to call the echidnas spiny anteaters, and the platypuses impossible.

We also learned at school that intrepid and unfortunate Captain James Cook discovered Australia in 1770. In the same breath, we learn of his encounter a week later with the locals. What? It's a bit like Edison saying he invented the motion picture system when really he 'borrowed' the glory from the French Lumiere brothers, or was it Edward Muybridge, or was it Joseph Niepce, who took it from countryman Leon Bouly, or was it Louis Daguerre?

To 'discover' Australia, Cook used the reverse route, down the east coast of South America, around treacherous Cape Horn, and westward across the oft unpacific Pacific. Usually, explorers and maritime merchants sailed down the west coast of Africa and went eastward around the Cape of Good Hope and on to the so-called Spice Islands. The gusty, westerly trade winds were useful, but the aptly named doldrum winds that were also in that part of the world were not conducive to rapid travel. Like a maritime highway, it was longer but quicker to sail further south than the tip of Africa, to more southern latitudes, then take advantage of the Roaring Forties winds to sail east, then turn north to hit Dutch East India. That is, if you guessed right and were lucky to head north at the right time. This alternative route was a bit like the roads in early human villages, always settled along river flats for access to water. Roads in the villages were subject to swamping, though, so better roads were eventually made away from the river on higher ground, but not used a lot because they were further away. In times of flooding and bogs, locals would avoid the low way and take the high way. The Roaring Forties was the highway.

As Dava Sobel points out in her spectacular book *Longitude*, knowing where and when to turn north was a bit tricky for sailors. In fact, it was pure guesswork. My smart mobile robots navigate by what I call parametric mapping using sensors, but my not-so-smart hobby robots navigate by dead reckoning, little more than estimating, exactly the same as explorers did back then. At times, presumably accidentally, they went too far and hit something entirely different, or nothing, perishing in the Great Southern Ocean. Abel Janszoon Tasman did that, not from the Cape of Good Hope but from Batavia, hitting what he called Anthoonij van Diemenslandt in 1642, but the same idea. In English, Anthoonij van Diemenslandt becomes Van Diemen's Land.

So, we were called Van Diemen's Land until 1856 when we became Tasmania in a bit of whitewashing. Back then, instead of Europeans being called Tasmanians we used to be called Vandemonians. With Tasmanian devils, and demon in the name, and ruthless penal history and colonial connotations, and efforts to eradicate the indigenous population, there can never be enough whitewashing. So, Australia, of which Tasmania is part, was discovered by Tasman in 1642, not Cook in 1770. Or was it someone else 60,000 years earlier, walking down through what has since become Bass Strait, stopping and staying at Land's End, thinking, wow, and calling it utopia?

Fast forward to 1950 and at Waddamana, a tiny highland village in the centre of Tasmania, is a Vandemonian called Allan Branch, a hybrid of all of that history.

I was born into a pretty average but troubled working-class family; rural, isolated, limited education, but innately an intelligent family, hard-working, hard partying, and with aspirations greater than our skill set would ever allow, most of us destined to achieve little. Which was my prognosis for life. In my 1959 green FB Holden one day, after doing some mechanical work in my Uncle Cedric Calvert's garage attached to a petrol station, in a rush I drove out backwards with my right-hand driver's side car door open, looking behind me over the other shoulder, and took out the concertina doors that were only open enough to let the car in and out, if its car doors were shut. Shattered glass and wood splinters went everywhere. Uncle Cedric was my dad's brother-in-law and his good friend, so I was in for big trouble. I drove off scared, embarrassed, unable to correct this massive damage to the garage doors. Later that night I drove carefully past the place and saw that plywood panelling had been screwed to the frames, so I knew there had been a temporary repair. But my dad was vicious, and it was a few days before I had the courage to face up to him and Uncle Cedric. My dad simply looked at me in disgust and said something that still echoes today.

'You'll never amount to anything as long as your arse points to the ground!'

Now, I knew a bit about anatomy, and nothing can be more motivating than knowing your arse is on the line. So, with Dad's coming-of-age instructions for success, I took on the world. Admittedly, I did not know initially that was what I was doing. Travelling blind, because no one in my family had taken on more than Tasmania before. Initially taking a decade to discover who I was, then two decades in heady Silicon Valley's advanced technology corridors, hobnobbing with the leading pioneers, then two decades in the upper echelons of the global corporate world. Coming out of it a little bit arrogant, a little bit unhumble, a little bit thin-skinned, but with my delicate anatomy unchanged.

It wasn't plain sailing. It wasn't a linear path like I've suggested, not all downhill. It was not a schuss, more like a slalom with the odd ski jump and side trips to the piste with the inevitable catastrophes. Some crashes were life-changing disasters.

CHAPTER 2

CAROL

It is a dark, brooding 'Hound of the Baskervilles' road, remote; anyone approaching is seen very early on. The headlights shone through the side window of the house for ages as the vehicle progressed along the straight, bumpy road, like hunters with guns and illegal spotlights, into the living room, creating bizarre shapes and shadows.

Only when the car stopped outside our place, a tiny gingerbread cottage at a fork in the road, was it obviously a police car, and presumably a problem to be avoided. My wife, Gay, opened the door while I hid in the back room, but within earshot, able to hear bits and pieces. I overheard something about death and instantly knew my dad, who was deteriorating from years of smoking and abusive drinking, had died. But I was wrong, and the shock of the truth was on the borderline of survivable. I wanted to die then and there, and it was certainly the end of everything about my life to that point.

'Can we speak with Allan Branch please?'

'He's not here,' she lied.

'Are you his wife?'

'Yes.'

'I have some very bad news; can I tell you and you relay it to Allan when you see him?'

'What is it?'

They talked for over an hour, framed in the doorway. I think they did not believe that I was not there, and she struggled to get rid of them. My wife, now ex-wife, was, is, a gorgeous woman and often had trouble getting rid of men, particularly those in uniform.

It was the only time I prayed and meant it. I mean really meant it … begging, pleading, imploring … and not just thrown out there as if to catch someone or something like a Native American dreamcatcher or an upside-down horseshoe, but to the actual God like He owed me something. I was close to Him. Not spiritually but physically. I was in the firmament, not God's haven, but 40,000 feet closer to Him, Ezekiel's vault, nevertheless. Like levitating, willing myself to be as near as possible in case that made it more likely to be heard. In that instant I hated Him but needed Him, and hated myself for needing Him.

Travelling south to north, flying as great a distance across land as it is possible if you are in Australia. Once before I had flown from Australia to New Zealand, my longest previous flight, and like with JFK, or the election of Obama, or the death of Di, I remember clearly being there in NZ when Elvis died. So that was 1977. I flew there to avoid a woman, and it worked, sort of. But even that international distance was not a match for this haul across Australia. This was 1981. This time I was flying to meet a woman. Out the window, country as sunburned as the poem, as red as the rumours, matching the tint of my rheumy-looking eyes.

My job was to fly to Darwin, in the northern tip of Australia, from Hobart, in the southern tip of Australia, to pick up my sister Carol and bring her home.

It had already been a few days, but I still could not stop the tears, those types of tears that simply leak out uncontrollably without the audible sobbing. Even now, 40 years later, with so much life lived since then, it was easily the saddest time in my whole life. Around this time, people very close to me were dying, but that could not match this pain. I was the eldest of six kids and Carol was the youngest by a decade. I'm supposed to die first. An unusual bond was the result.

Apart from a connection in Melbourne, unavoidable because all flights from Hobart at that time went via Melbourne, there was a transit stopover in Alice Springs, the centre of Australia. Over the years I have lived and worked in some of the hottest places on the planet – Texas, Cairo, Singapore, Townsville, Hong Kong, Bangkok but the town called Alice on that day was as hot as anything I have ever experienced. Not that I had any interest in anything other than my immediate task, no desire to visit or explore the place. After walking outside to stretch my legs and immediately back in again, I sat catatonic in the simple unair-conditioned waiting area at the simple airport, keeping cool by not exerting, anxious to reboard and continue.

In the waiting lounge I knew people were looking at me, sympathy in their eyes and expressions, knowing I was in the midst of tragedy but not knowing precisely what. Wanting to comfort me I could sense, but not knowing how. I didn't care that they cared.

Then we boarded and continued. Ayers Rock, as it was still called then, at least by people my colour, was visible out the left of the plane, but I was on the right side, and did not see the giant red or orange or brown or yellow boulder until I was still on the right side on the return flight. I might as well have been on a routine bus ride from Hobart to Launceston. Not napping, just looking, not really thinking or seeing, staring blankly. The world had changed for me even though outside the double-glazed plane windows it was still the same for everyone else.

In Darwin, the weather had moved from hot-dry to hot-humid, and my wet tears continued unabated, unembarrassed. I was met by my brother-in-law Michael, sheepishly, warily, devastated. He had flown back to Darwin from Adelaide, where he had been with his two infant children to stay with his parents while my sister was unaware in hospital in Darwin. It was Michael who called me and told me the full story not long after the police had left that day.

We stayed at his friend's house, me in the living room on a flat mattress on the floor, and which I carefully centred under the ceiling fan which I never took off high. Darwin was, is, a small town, no high-rise buildings then, clean, green, slow. I remember nothing of it except that it was expensive. It was said that people, like my sister, go there to escape something, and can never save enough to escape out. Clearly, that's not true; she found a way out.

Darwin was famous as the entry point of Japan's World War II attempt to invade Australia. Normally I'd have explored and investigated, but I have no memory of being interested, except I remember the tiny local gaol being called Fanny Bay, because my great grandmother was named Fanny, not Francis or any other variation. A word with different meanings in Australia and the USA.

Darwin was a backwater. In 1974 it had been hit again, that time by nature as Cyclone Tracy pelted the town with 200-kilometre-an-hour vortices over three days at Christmas, killing 71 people and causing a billion dollars' worth of damage. Almost as much as the cyclone of 200 Japanese war planes when they invaded in 1942 straight after they had pelted Pearl Harbor. Beaches between the Darwin harbour and Fanny Bay were home to the deadly bluebottle jellyfish, formidable weapons fittingly also called the man-o'-war, making impossible any thought of swimming in the hot tropical water, not that I had any intentions. Why anyone would go there I do not know, and it is little wonder that it's the beer-drinking capital of Australia.

My brother-in-law's friend was a private pilot, small planes in Papua New Guinea. A few years later I heard he was killed when his plane crashed navigating the New Guinea high mountain weather on a commercial contract.

The next day I met my sister's friends, my first encounter, as far as I know, with lesbians, open and proud, and as sad as I was. Several sharing a house, hippy style, which was the nature of our meeting. Some of my sister's things were there, mostly music, and I later purchased a copy of the record they said was the one she played most of the time. Steve Forbert's *Jackrabbit Slim*. I played it to try to put my head in her space. Then I played it because I loved it, as I loved her. I haven't played it lately, but will today or tonight now that the memory has resurfaced. I think 'Romeo's Tune' on the album is the one.

It was also the first time I was hugged, as far as I know, by lesbians. It felt no different from a hug by anyone else. It was lovely. It was heart wrenching. But worst was to come. Carol was not a lesbian, and I still do not quite know it all, but I think it was just her best friends sheltering a troubled soul. She was certainly a feminist and an activist. Just a year before, the last time I saw her, she visited me in Hobart, proud to have been selected to deliver a speech on feminism at a national conference at the Wrest Point Hotel, Australia's first casino. Proud to make it visible to her big brother, who usually dominated the spotlight. I'd give anything to replay that visit, to hug her and never let her go.

The next day we met with Carol's official counsellor. It is not healthy for a counsellor to get too close to a client.

'I'm sorry, I should not be crying. I am supposed to be the strong independent one.'

'That's okay,' I said, 'it is sort of comforting to see that Carol had such a strong effect even on you.'

'There is something crucial I have to say,' the counsellor said between her tears. 'It is important that you do not blame yourself for what Carol has done.'

How could I process that? Of course my mind had been racing around all the times we had been together, searching for clues, hints, suggestions of what was ultimately to come. For what I could have done or said differently if I had only known. Of course it was my fault. Her big brother was supposed to protect his little sister.

'I can't help it. I am her older brother and she is the youngest. I feel responsible. We had a bond.' Present tense, no acceptance that she was gone.

'Carol was a troubled soul. What she did was because of her, nothing to do with anyone else. She mentioned you many times, was proud of you, and there is no way you can take on any burden of responsibility.' Now the counsellor had gathered her composure and was sounding like a caregiver.

My sister had attempted suicide a few times, so there was an appointed person to look after her, look out for her. Suicide is in my family. The experts say it does not run in families, there is no genetic basis to it. That makes sense because it would evolve itself out. They say it is a learned thing, from books and movies and anecdotes perhaps. A misguided solution to an impossible situation. Unbearable situation, in the case of my sister. One sees someone else in the family do it or try it, so they do it or try it. So, it seems to run in families. It seemed to run in my family: an aunt, two aunts, a sister, two sisters. I think alcohol was the catalyst. Leading to failed jobs, failed relationships, failed dreams.

Carol's cries for help were unknown to me, but centred around her marriage. She was admitted to hospital in Darwin, and on release, instead of being able to rejoin her husband and children, one year and three years old, babies, they had disappeared to Adelaide, and instead of happiness her sad life was unchanged. They also say it is a cry for help. But it is always the survivors who cry. I'm the one crying. I'd never cried so much.

Even before booking the flight, but knowing that I was to fly south to north, I was already praying. I mean really praying. Every muscle clenched as if the force and tension and effort would push the prayer out. And here is the awful thing. I wanted it to be someone else who had died. Not my sister, anyone else. I was praying for God to have killed anyone other than my beautiful troubled, 21-year-old youngest sister Carol. The sister who had fought so hard for life in the beginning, born three months too early, at a time when three months was too much and they usually died. But she did not die. In the early years she had yellow baby teeth from some wonder medicine they gave her to make her unformed lungs breathe. I remember that. Carol, the closest to my mother in stature, petite; in temperament, defiant; in personality, evasive; in looks, beautiful.

Then there is the other matter.

There is nothing harder than crying so uncontrollably but trying not to make a sound. The same tears that were on my face on the plane a few days later had started with the police on my doorstep, the ones who had been told I was not home. There I was struggling not to be heard, howling silently. I remember being unable to function for crying for two days. My brother-in-law then called from Adelaide. Why not Darwin? I could not speak because of the crying and within minutes, as it sunk in, I hung up on him, unable to hold the phone, as if it weighed a ton, or was toxic. Slinking to the floor because my legs were not strong enough to stand.

From Darwin I flew back to Adelaide with Michael, met his parents for the first time. From the plane window watched Carol being loaded into the hold of the plane, then back to Hobart, inviting Mike and the kids to join me there, stay with us. In Hobart we buried Carol, and my mother screamed like a banshee. Mum's divorced husband got drunker.

For a long time I was sad about my sister, then I was angry with my sister for a longer time, then finally I understood that she was ill and did not intend to hurt any of us. But I cannot dwell on it too much, the tears still come too readily. No other event in my life does that to me. Now it is all sadness again.

Julian Barnes, in his heart-destroying *Levels of Life*, aptly describes recovering from grief as not like a train emerging from a dark, depressing tunnel to green pastures and sunshine, but like a seagull emerging from an oil spill. That is it. Grief ends, but life is not the same. As a shattered vase, glued back together, cracks visible, not as robust, quick to break again, points of weakness, not quite whole.

My prayers never came true, and I have not prayed since.

CHAPTER 3

TURTLES

Tasmania has no indigenous freshwater turtles. We are instructed to report one if we discover one because they are an environmental threat, potentially introducing diseases and interfering with natural habitats. I had never seen one. So, it was captivating when I lived in Townsville in the tropical north of Queensland and watched hundreds of turtles vying for sunny space in the green pools of the ever-flowing Ross River. I loved learning about animals and nature at school.

For reasons unknown to me, school was a breeze. Both of my parents had minimal secondary schooling but were innately smart. I think. Dad, in particular, helped me with my education in the early years, not usually in a way that added marks to my exams – the names of all the parts of a house frame for instance, or how to vote – but sometimes. When I was about seven, Dad showed me the simplicity of multiplying by 11. At school the next day, to show off, I mentioned it to the teacher, and she tested me, right up to 10 x 11, which I did not know because Dad only taught me up to 9 x 11. The simple ones. I should have learned then that it is the exceptions to the rule that provide the greater insight to how things work. From about grade 5, when I was 11, I managed to come first, second or third in every exam. This was despite being chronically asthmatic and missing long periods from class. I don't remember any transition in my thinking or change in my study habits. I remember being somewhat stunned by it all, particularly the first time I was first.

Chronically shy too. The scariest thing we had to do in class was to stand in front of our classmates and deliver what was called an oral – a short, spoken discussion on a topic. Often of choice, but sometimes set by the teacher, without using notes. Sometimes without notice, which was the worst, and sometimes the next day, allowing some thankful time for preparation. We all hated it and were all bad at it. Dad had bought me the *Australian Encyclopedia*, a two-book collection that enthralled me. From it came my choice this day for my oral presentation. Fittingly, it was the voice box, the larynx. I always studied alone, so had never heard the word larynx verbalised before, and with such unfamiliar spelling, the inevitable can be anticipated. The next day, I walked tentatively to the area in front of the blackboard, and in front of everyone said I would talk about the larynx, but I pronounced it 'larexyn' or some such mash.

'What?' asked my teacher, Mrs Hamilton. She was lovely, but elderly and hard of hearing and probably thought she had misheard. Mrs Hamilton was the only one we never feared getting the cane from. Her cane, a willow sapling, was old – maybe she had it since graduating – frayed like a cat-o-nine-tails, and painless across our legs. No one ever told her and we all faked screams so she'd never find out. But, of course, she knew.

'The larexyn,' I repeated, less confidently.

'The what!'

'The larexyn.'

Silence, and clearly confusion, as I went redder than the ochre that painted the bricks in the open fireplaces in the corners of the ancient, dismal Glenorchy Primary School rooms. Painting the fireplace and washing the blackboard were regular chores allocated to students. Remember blackboards? Remember ochre? I knew the word 'ochre' despite its unusual spelling because my cherished nan coated the fire bricks in the hearth with it on a regular basis.

'What on earth is that?' she asked.

'The voice box.' Not very sure now.

'Oh. That. It is pronounced larynx,' Mrs Hamilton offered helpfully.

Anyway, I remember getting 8 out of 10, a very high score, and I wonder now if part of it was for entertainment value. But it happened again with an impromptu talk on synclines and anticlines. In my career, presenting in front of audiences has benefitted from that nerve-racking early experience. The best presenter in our class was John McDonald, so well spoken, so clear, such interesting topics, so confident, the envy of all of us, and it is no wonder that he became initially a newsreader on primetime network TV and later worked in the media and film-making industry.

I should have learned a lesson, because years later with my very intellectual team in the car I drove past a local landmark building and remarked that it had an interesting façade, but I pronounced it far-cade like arcade. It took a long time for them to stop laughing and for my ochre face to blanch, and I have never used a word that I have not learned the pronunciation of again without checking it first. Being self-taught comes with perils.

By the time I left high school and started matriculation college, things changed. That was the first time teachers did not chase you up if you missed a class, the first treatment as a self-responsible person. Unfortunately, instead of self-discipline I discovered cars and girls. My best friend's parents took a trip back to Italy, and I taught myself to pick a door lock so we could use his family's new EH Holden, supposedly safely stowed in their garage, to cruise throughout the summer with our girlfriends. By the end of first term, I had bombed out and started working.

Before that, during the last year of high school, I needed some spending money and walked the streets of Hobart calling into shops and businesses asking for school holiday work. Victor Bergfield, destined to be my lifelong friend and mentor, gave me some tasks in his dusty old radio and TV repair shop on Elizabeth

Street, to become a parking lot and now a commercial building. By then I had developed some electronic knowledge from weekly activity sessions with Bill Dick, our science teacher, who had returned from an Antarctic expedition as their electronics engineer. It was quite exciting to have a teacher fresh from Antarctica, with all his stories and the intriguing bits and pieces of scrapped electronics he brought back. After my unceremonious departure from college, Vick offered me a full-time job for $8 a week, which became my first official job. He was amazing. It was the days of vacuum tubes still, transistors only slowly becoming more and more established, and no such thing as an integrated circuit. Vick showed me how to fix and repair things no one usually knew about, like voice coils in speakers, or graphite coating on potentiometers, or the windings of IF cans. It was very different from Mr Dick's rigorous engineering style. This hybrid electronics experience was something I rarely found in others and became invaluable for my career. Homegrown electronics repairers didn't care too much for theory and rigour, just anything that worked, and contrastingly engineers never broke the rules.

My fledgling electronics skills and interest were supplemented by some mechanical aptitude, having bought my first car, a 1948 Morris Minor, which only ran if I maintained it - all the time. So, I was okay with electromechanical devices. More so than Vick it turned out. In the front of his shop where all the large television sets and radiograms were kept was a jukebox. Quite exotic and out of place. It was waiting to be shipped to the Australian mainland to the factory where it was manufactured, because Vick had been unable to repair it for the owner. I asked him if I could look at it, and as there was nothing to lose, he allowed me. Within a few minutes I had it working, a pair of fused contacts on a relay. No one was more surprised than me, and I was Vick's hero from then on. We were each other's heroes.

One of our customers was Peter Muir, who came in a couple of times to collect his expensive high-end stereo system. The first time Vick was not there, and I could not help Peter, but we got talking. He had seen the broken jukebox and noticed it was working, so was intrigued when I related the story. The next time he came in he mentioned that there was a job for a junior technician in the state department of agriculture where he worked as a field technician. He had mentioned me to the boss, Ron Richards, and they wanted me to take the job. Apparently, I had made an impression with Peter and subsequently with Ron. Ron had to pull some strings to get me without advertising the position, but that was my next job.

All of that electromechanical experience came in handy. Another life-long friend and mentor was a Rasputin-like German immigrant called Eddie Sauer. His name was Kurt, but Kurt Sauer was too unsavoury for English ears, so Eddie it was. He was the science laboratory manager at Cosgrove High School, where a privilege was to be appointed by him as a lab monitor required to clean all the glassware and help prepare the chemicals for class experiments. It came with special perks because we could boil water in a beaker and make coffee at lunch time. A perc perk. Almost no one drank coffee back then, so we did not know the word percolate. Eddie, true to his mad-scientist image, had developed a way to pump superheated water into a car engine to make it run on only 20 per cent petrol, and it was a hobby passion for both of us then and after I left school. On his death, years later, his wife Marie referred to me as his best friend.

Eddie had acquired racks of what are called unit-selectors from the old Hobart telephone exchange and built the school's timing system from them. They were a solenoid-driven relay device with an arc of step-wise contacts, cycled through a pulse at a time. In the old rotary dial phones, each time you heard a ting as the dial rotated back, it was pulsing the unit-selector back at some exchange. It was my experience with these curious devices that built my knowledge of relays and relay blades, leading ultimately to the nickelodeon repair. I had some banks of these devices and built a universal watering system for all the glasshouses at the horticulture research centre where I worked with Ron. It would turn on watering systems at various times, with differing cycles for day and night, summer and winter. I was 17 years old and had no sense that this was anything but normal.

Days were spent on field experiments, injecting orchard tree limbs to determine nutrition inefficiencies, or into the ground to correct missing trace minerals, implementing hormonal thinning on orchards to even out annual yields and introducing cell packaging to get apples to England on ships without bruising.

Lunchtimes, my adult co-worker Barry Russell taught me how to do cryptic crosswords. He also taught me a nonsense ditty I have never forgotten: 'Root Toot, Toot Shoot, Riddort Fort the Fluff Duff, Rave Dave.' Barry also taught me another ditty, more pithy, which I have in turn taught to kids of parents where I have been a house guest around the world, who now as adults repeat it back to me with glee when I visit.

'One One was one racehorse
Two Two was one too
One One won one race one day
And Two Two won one too.'

The masses of field research results had to be assessed. All field measurements were evaluated with rows and rows of statistical tables, all done tediously and meticulously by hand. Ron taught me about variance, standard deviation and Student's t-tests, and about accuracy because it is a lot of hand calculation to redo if mistakes are made, no calculators then. Ron showed me another skill I've rarely seen since. He could glance at a page of data and spot an error as if it were illuminated by a lantern. I can do it with financial reports and research papers and do not know if I've learned it or if it is inborn.

When I was 19, another job came up for a laboratory technician, this time in the Department of Surgery at the local clinical school at the University of Tasmania, a part of the Royal Hobart teaching hospital. At the time it was the only teaching hospital for student doctors in the state. The head of department, Professor Robert

Mitchell, was a pioneer kidney transplant surgeon, which is the research we did in the labs, that and open heart surgery. I became a laboratory technician, an operating theatre technician, a research assistant, an animal house technician, a skilled teaching-aid producer and more.

Vick and Eddie were European, sophisticated, worldly. Barry and Peter and Ron and Robert were graduates, professionals, different vocabulary, different tastes, different deportment. They were not tall or broad, did not speak with deep gravelly voices like a stereotypical macho image. They were assuredly softly spoken. Nothing like the individuals I'd encountered in my past. My family and family friends were working class, even uncouth. The exposure was broadening my horizons, breaking the cycle. Their self-assurance and confidence in other people meant they had no qualms about letting me have a go at a jukebox or a car engine or a watering system or a heart-lung machine.

There seems to have been a three-year-itch pattern to my jobs. There still is, which makes for a curious curriculum vitae. There were a couple of extra-short jobs: one with a jukebox company, coincidentally, and one as a nurse trainee at the Royal Derwent Hospital, an asylum when such things existed. But in the midst of that, my political hero, Gough Whitlam, became Australia's idealistic prime minister, and as well as getting us out of the Vietnam War he made tertiary education free. Hoorah for ideals! So off to the university medical school I went. There was no possible way anyone with my poor, working-class background would have ever gone to university otherwise. Interestingly, while I focussed on physical sciences at high school and had no interest in biological science, a look at my jobs shows that they were mostly biological in nature, hence my choice of medicine.

At one point before this, apropos of nothing, or maybe because the biology and wet science was starting to excite me, I undertook a night class and secured a high distinction with college biology. One part of the course was an essay that had to be about a personal experiment. How any student could have yet experienced an experiment I felt was unlikely and a bit unfair, but I had a plethora of them. In the agricultural field experiments, I learned that copper inhibited the uptake by roots of other nutrients from the soil, so I wrote about that. The lecturer, chubby and affable, called me out in front of class and said in all his years he had never had to check the veracity of the essays but he had to delve into mine and found that it was indeed a fact. With that return to schooling I went back to college and in one year collected physics, chemistry, maths A and maths B for a five-subject matriculation certificate, which was needed for entry to university. It was not my first introduction to scientific method. Unbelievably, and illustrating the difference in education in Tasmania back then, my fourth grade teacher, when I was ten, performed a class experiment showing that vegetation prevented erosion. An unbelievably prescient choice.

Uni was also a three-year itch, and I departed after learning the basics of psychology, anatomy, biochemistry and physiology, so three more years went by. Then another three. By now I was married to Gay, had been to New Zealand and met Bob Duncan, an entrepreneurial businessperson in Auckland. I worked in his electronics repair and sales business called Bear Electronics, but on returning to Tasmania from New Zealand I needed a job. I sat an entrance selection exam with a thousand others for a federal government position, was called immediately and took a job with the Australian Federal Government. Nothing to do with science, technology or medicine. I was promoted quickly to the second highest level and charged with powerful control officer and productivity officer duties. It was my introduction to finances and performance, large organisational structures, bureaucracy. A year later I was offered the directorship at a small government division in Katherine in the Australian Northern Territory, an unbelievable top cat opportunity that everyone urged me to accept. But something got in the way to change my direction again.

While in New Zealand in the 1970s, between all the parties and the smoky Māori hongis, and mixed with the music and laughter, I discovered a level of inventiveness I had not experienced in Australia. It was driven by New Zealand's poor balance of payments at the time and the consequent ban on imports. If you wanted something, you needed to make it at home. The first 4-byte microprocessors were just becoming available. One of our engineers introduced me to them because they were in some video arcade games, so my electronics skills went up another level.

As mixed up as my working life had been so far, the experiences I'd had in state and federal government, large and small companies, a university, a stint overseas, and a variety of industries led to an inordinately large exposure to organisations with their various structures and entrepreneurialism, which not only served me well for what came next but is still the basis of what I do today.

Bob Duncan, big and bold and brazen, nicknamed Bear, hence the name for his business, kept in touch from Auckland after I returned to Tasmania and came over for a visit at times, staying with me and Gay at Lonnavale. On one visit he learned that a local electronics retail shop and service centre was available for sale. He suggested we buy it. 'How?' was my question and he explained that the business was insolvent and we could get it by taking over the debts. The business was Aero Electronics. At this time I was still working at the Federal Government Department of Social Security, and it was also the time my beloved Nan passed away. So, I quit my lush job and together we bought Aero, a brightly visible, street level, retail shop with full glass window frontage and dusty workshop out back. We set up in business as an electronics reseller and repair shop. My friend Vick Bergfield, who gave me my first work when I was a student trying to earn some pocket money, and then gave me my first paid job when I left school, came to work with me as my manager. With our combined skills we advertised to fix much more than just radios and TVs and got lots of unusual equipment to repair. It was the last place here in the shop that I saw my sister Carol alive when she called in to tell me about a seminar she had flown in from Darwin to present in Hobart.

A dear friend from my matriculation days at Elizabeth College, Sandra Wills, stayed on at the college after she graduated, and with teacher and IT expert Scott Brownell they were pioneering the use of a new computer software teaching language called Logo. Tasmania was the first Australian state to make education compulsory and had a solid education history, so it was no wonder that one of the world's first versions of Logo was written by this team and being tested at the college. Pioneering personal computers like Atari, Apple and Commodore were just coming in, so it was also the earliest days of the PC revolution. But their Logo was written on what was a more powerful computer at the time, a Digital Equipment 16-bit PDP11 minicomputer (not a microcomputer, which is what a PC is).

Despite being designed to be an easy language for kids to program, with intuitive instructions, Logo was a powerful piece of software. It had all the capabilities of the more advanced languages of the times, like Pascal, Algol, Fortran, and was superior to the basic programming language that was now available on most of the new PCs, called, surprise, BASIC. It had recursion, subroutines and things called objects, for instance. Logo had been created by Professor Seymour Papert at the MIT Media Labs in Cambridge, Massachusetts. Papert was a renowned child developmental scientist, a student of Piaget, a media expert and a pioneer in artificial intelligence. He was every bit the quixotic scientist – Jewish, South African born, fiercely aloof.

What was unique about Logo was that it operated a turtle, and Tasmania had none!

Students using Logo could instruct an animated screen turtle to move around with simple commands like forward 10, right 90 degrees, forward 10, stop. The turtle's trail was drawn on the screen as it moved and the student could create shapes, designs and art.

The whole system worked best, though, if the computer and its Logo language could drive an actual turtle, a remote robot, a physical device, with a pen to draw as it moved around on paper on a tabletop in a classroom. With a real piece of hardware, students would firstly be motivated because they were controlling a robotic device, and secondly, such a robot moving around and drawing as it travelled left a trail as evidence of the computer instructions the student gave it. So, for instance, if the student wanted to draw a triangle, and they calculated the 180-degree internal angle rule of a triangle incorrectly, the final side of the triangle would not meet up where the drawing started. The instant visual feedback helped the student to self-learn. Students using real turtle robots would often have to be kicked out of class because they stayed behind to play with the robot, all the playing leading to learning, the purpose of the Logo language.

Seymour, together with an outside company, had built such a robot, called the General Turtle, and Sandra had one.

It was extremely expensive because it was overdesigned – it cost something like $4000 in 1979 – but it was a key feature of the research they were doing at Elizabeth College. One day Sandra came into my shop at Aero Electronics and asked if I could build one. Their robot was complex and unfriendly, and, not surprisingly, was not working. She asked not only if I could build one, but if I could build one that wasn't so expensive, so that classrooms everywhere could have them. As a device to be handled by hundreds of students, it also had to be robust and easily maintained. Being who I am, of course I said yes, I can build one, and the Tasman Turtle Logo Robot was born, – the only indigenous Tasmanian turtle.

It was also the genesis of my collision with the high-tech sector dominated by Silicon Valley, a fast-paced, global, competitive, survive-or-die world.

While I was designing the first Tasman Turtle for Sandra, I started receiving orders for them from her promotion of it in educational circles around the world and slightly later by the promotion from Australia's Federal Government export trade department. Sandra needed the first Tasman Turtle robot to present at a conference and so precious was it to her, that she flew with it on her lap to the Australian mainland. That first one was acquired by Wollongong University, where later Sandra was to be a full professor. It looked a bit like a domed cloche that you put over a dinner plate, a half sphere, except all transparent so you can see the innards mounted on a flat transparent base.

So, from Aero I incorporated Flexible Systems to manufacture and sell Tasman Turtles worldwide. Flexible Systems was Tasmania's first high-tech company in the 'Silicon Valley' sense. I brought in two partners, Adrian Firth and Len Whelan. Adrian was a successful local businessman doing lawnmower repairs, so had business and mechanical knowledge; Len was interstate in Victoria and had a company manufacturing simple pick and place robots for factory automation, so was already into robotics. These introductions and corporate consolidation were initiated by state and federal government officers, aware of the potential for this new startup, but equally aware of my inexperience in anything except the technology, especially business and management. It was a smart move on their part and without it we may not have been as successful.

Thousands and thousands of our robots were manufactured in a minimalist upstairs factory suffering the fumes of noisy two-stroke lawnmowers down below, in the heart of Hobart, Tasmania's capital city of around 200,000, to be shipped into classrooms around the globe.

The Tasman Turtle was the world's first successful commercial robot. It was not limited to classrooms. It was pioneering. We discovered that universities and, surprisingly, even technology companies bought them for their own research or to reverse engineer. The 1979 Tasman Turtle robot had speech synthesis and could talk to you – speech recognition – so could be controlled with a microphone, had touch sensors to interact tactically with its environment, had a few fun things like lights for eyes, a solenoid operate pen in the centre to draw or not draw its path as programmed, precision stepper motor wheels for accurate movements, and a clear semicircular shell with transparent base so that its drawing could be seen through the robot body. It was driven by our own version of Logo sold with the robot, all for about $300 in its simplest form.

Not long afterwards, in Boston, Adrian and I visited MIT, the birthplace of Logo, then went across the way to Harvard to guest lecture and demonstrate it, then to meet Papert himself, who was in Quebec, Canada. When Papert visited Australia for a lecture tour and we met up again, he wrote about his shock, and delight, when he went to classrooms in Australia to discover students everywhere using Tasman Turtles, dozens and dozens of them.

The name Seymour Papert was already known to me. When I was at Elizabeth College earlier as a student, I entered the national students' science talent competition with my work called 'A Trichotomy on Vision', winning a place at the state level. One third of my work was an attempt to reproduce the 1959 work of Jerry Lettvin and others, also from MIT, in their ground-breaking paper 'What the Frog's Eye Tells the Frog's Brain'; but another third was an artificial intelligence network performing visual pattern recognition that could read handwritten letters with size and other invariances. It was not a hardware circuit but a simulation, which a few years later was surreptitiously programmed on the University of Tasmania's administration computer while I was a student at the medical school. The circuit was a modified neural network.

Just two years earlier, Papert with Marvin Minsky, two of the world's premier artificial intelligence pioneers, stated incorrectly in a controversial and now refuted 1969 book called *Perceptrons*, which is what neural nets used to be called, that a neural net could not be a useful multilayer pattern recognition vision solution. Luckily, I ignored that perceptron perception.

The more interesting work had been by David Hubel and Thorsten Wiesel on cat's brains and identified the features detected by the primary visual cortex of their brains. My neural network circuitry was really a receptive field feature extractor, which allowed the invariance. In other words, it would recognise a shape (a letter of the alphabet) of any size and in any location in the visual area and white on black or black on white. I wish I had met these scientific giants. Another Nobel Prize laureate, Erik Kandel, made an equally significant discovery about how long and short term memory is accomplished at the synaptic level. When mentioning this to my best friend Zack Rosenfield in NY he said Erik was a neighbour. Zach offered to introduce me, something I never did and still regret.

Interestingly, I used to vacation on Deer Isle in Maine. Seymour retired there nearby at East Blue Hill for the remainder of his life. Although living within close distance of each other we never met up again.

CHAPTER 4

FIRE

We camped a lot – holidays, long weekends, as often as we could. Rough camping, no luxuries, sand and surf, dust and dirt, smells and smoke from campfires, salt from the ocean. A remote stretch of hidden beach on Tasmania's east coast. Appropriately called Friendly Beaches as if beckoning us. Explored originally by Louis de Freycinet, who managed to hide his wife Rose on board illegally disguised as a man so they could be together; so very French. Now a regulated national park bearing Freycinet's name, controlled, limited access for a steep fee. Then, pristine, unknown, uninhabited and impossible to find, partly because we removed the signpost to the insignificant dirt track from the main road whenever we went there, to deter others. Then, no entry fee, we were Tasmanians and this was Tasmania, our land our rights. We had tents, bikes, camp chairs and tables, wind surfers and beach buggies, surfing and diving gear, guitars and food, camp lights to party as late as we could stay awake. So, perhaps not *that* rough. Three or four couples and a bunch of kids each trip, friends or family. We did this for about a decade through the mid-1970s and 80s.

This particular time we were in the middle of winter, though it was still mild and good for camping, when a police car came up to our camping spot, a flat grass-covered clearing surrounded by stunted salt resistant shrubbery, near the edge of the sand dunes. This was a first, to have the police visit. We never did anything illegal, except remove signposts of course, and each time we left it was like we had never been there. A family rule that had been instilled in us as children. Also instilled into my nieces and nephews from early on; including my sister Carol's kids.

The police asked who we were and asked if we lived at Lonnavale, a remote rural village an hour south of the capital city Hobart, too small to be a village, too tiny to be anything, a collection; not even a collection – a couple or four of farmhouses. As far towards the protected Southwest National Park of Tasmania as you could go before hitting the Snowy Range. That was us.

'We're sorry to have to advise you that your property at Lonnavale has been destroyed by fire,' was the shock delivered that day, leaving all of us embarrassed and speechless. The police, unable to look at us, sorry for us.

Cutting our camping holiday short, slowly each of us realising that our lives were about to change dramatically and forever, we were stunned into silence. There was nothing to say. Long before we saw the devastation of our home, we already knew what this meant.

We scattered to pack up, using none of the care from the trip out, and in our respective vehicles drove unusually slowly back to the raised ruins of our beloved farmhouse. No one spoke the whole way. There was none of the revelry of the trip out a day earlier – meeting up at a prearranged place, vehicles crammed with kit and kids, chasing, racing, passing each other in the excitement of the imminent adventure, stopping for snacks or at the unmatchable sea-swept vistas en route.

The absolute destruction of the property was difficult to comprehend when we arrived. Ashes to ground level. Chimney stack like a sentinel. Twisted corrugated roofing iron charred and caught in agony like Pompeii victims. Nothing. One memory was my coin collection of Australian copper pennies, a coinage recently replaced by new decimal currency, becoming collectible, now worthless melted alloy, shapeless like a miniature Henry Moore sculpture. Not even that, a metal cast of a cow pat. Nothing was salvageable, residual or even recognisable. It was total. There was no reason to stay and we made separate plans for somewhere temporarily to live.

Less than a year later, Gay and I were living next door on another small hobby farm at Lonnavale, literally the next block from the one burned down, which we bought on a whim after being burned out. It was an opportune private sale, and we paid the owner a monthly instalment over a few years, about three from memory, the only way we could get a new property with no insurance payout from the fire. We were insured but the company never paid us. It was the same cottage house where I learned about my sister Carol, but that was later. This day the white wall phone in the hallway rang and we received a call from a young girl in a neighbouring farm up the hill. She was panicking and scared because only the young ones from a family of seven were there, and they were cooking on the stove when the saucepan of oil had caught fire. They did not know what to do and could only think to call me, the nearest adult. I raced up to their house, only a minute or so away, into the kitchen and saw the flames. They had a pot of water on standby and I yelled, Don't throw that on it! I grabbed a thick woollen kitchen towel hanging nearby and smothered the fire, the flames disappearing instantly into a puff of smoke like a conjuring trick. The tiny girls were amazed. I explained that water is never used on oil or petrol fires, only with things like paper or wood. Water would have spread the fire like, well, like wildfire, and the whole room would have been irreversibly alight. The pan was warped, the top plastic control knobs of the electric stove were melted, but otherwise there was no damage, but it showed how close another house fire might have been.

Tasmania is part of Australia. Greener and cooler, yes, but dry and hot in summer and just as prone to fierce bushfires like anywhere on the mainland. Propelled by the incendiary eucalyptus oils. I was familiar with bushfires. A decade earlier, in

1967, the encircling flames that almost entered the capital city, killing people and burning suburbs, left similar scenes to our burned house. In 1967 we roamed around afterwards, awestruck and dumbfounded by the remains of people's homes, dreams, smouldering toys, scorched cars … of people's lives. But it had no real personal impact until it happened to me.

Not even when it happened to my widowed Aunt Molly with her three children, who were living on the mountain side in the midst of that 1967 fire in a black-oiled, vertical board house provided to her deceased husband as part of the war effort for veterans of World War II. The fires were fierce and my equally fierce dad acted. On hearing radio news of the fire's progress to the Hobart suburbs on the mountain, he typically ignored a police barricade and charged through to find them hiding terrified under their house. He brought them back to safety, thumbing rudely past the same police cordon on the way back, to say, 'I told you!' I had been in class at college and they let us all go, saying get home quickly. As I bussed back home through the suffocating smoke, not medicinally aromatic like eucalyptus usually is but acrid like burned chillis, that enveloped the city, I had no idea of this rescue taking place at the same time. The next day we saw that my aunt's house was also razed to the ground. They'd have perished, so he saved their lives. Insurance built her a new, better, yellow weatherboard house on the same double block of land. Everywhere, the chimneys were tombstones marking where people had lived.

Many years later, again on the mountains and shores and outskirts of the same city, a firestorm encroached on the nearby seaside town of Kingston, where a single elderly female friend lived. A talented musician, spinster and owner of two of Mim's dogs. Mim was by now my new partner, and we had been together more than a decade. I cannot recall if Mim reacted instinctively just because she was concerned, which would be in character, but I think she had an emergency call to help her friend because her back boundary fence, cinder-dry wooden palings, was on fire, smouldering at the base where flying embers had blown in on the fire draught and become trapped.

Saving dogs, any animals, at times of fires was a priority for Mim, something she had detailed plans and arrangements for back at our equally susceptible rural property at Wattle Hill. We were miles away from Kingston, on the other side of the Derwent River, the second deepest natural harbour in the Southern Hemisphere, maybe an hour's drive. Without thinking twice, we jumped into our car and took off. Across the mile-long Tasman Bridge over the river, we knew from the TV news that the shorter main highway was blocked, so we tried the alternative route around the old winding coastal highway, but the fires were too intense, and not being my father I had no interest in charging past the police barricades. We turned around to take the only other way, the long way up the mountain through alpine backroads, and again met walls of fire. With little choice this time and no services stopping the non-existent traffic (no one would be silly enough to head into the direction of the fires, would they?), we pushed into the smoke and haze and flames and heat and flying burning debris. Not sure if we would get through or be incinerated, but managing to get to the other side, on to our friend just as the fence fire was moving across the yard to her house. At almost the same time the fire rescue service arrived and took over and all was okay. All that remained when we checked later was a black singe from the fence across the lawns touching the rear wall of the house, showing how close a call it had been. The Tasmanian volunteer fire service saved every house that time.

Little wonder that so many kids want to be firemen, the most heroic and daring profession they could imagine. More than police, more than pilots. More than just dreaming, kids want to be brave and daring. So many individuals recognise it as a worthy vocation. In Tasmania, the volunteer and paid recruitment lines are constantly oversubscribed. Nevertheless, there are always openings because it is very selective with a rigorous and taxing training program; it's the intelligence and psychological profiling ensuring rapid decision-making when there is little information, and calmness under pressure, that screens most out.

Back at the ruined house at Lonnavale that day, the comfortable living arrangements we had in our family commune ended. We all went separate ways, never to live close together again, physically or emotionally.

The previous year, as young adults, all struggling financially, we held meetings, discussed everything in a sociable manner that is impossible now, deciding to buy a large property together, live together, pool resources, save money, share costs, in a democratic structure at Lonnavale. We bought it in January 1976. We had been there only a half-year when, like all communes, all democracies, it rapidly fractured, with four of the nine of us quitting within weeks. We were not unlike our parents, uncles and aunts, friends and hangers-on, and their drinking parties, which as kids we loved. Starting well enough, watching the music, dancing, laughing, playing until time for us kids to go to bed, but then later loud crashing sounds and shouting heard from our bedrooms, sometimes waking me up. Seeing black eyes and sore expressions next morning. Cuts and bruises and miserable silence, typical of a battle aftermath. All subdued, hungover, all forgiven or more likely not remembered, and having no discouragement on the next party to come.

We proved to be no better at sustaining friendly relations.

Among our ashes, the loss of expensive furniture and electronics, televisions, stereos, furniture, guitars, appliances, material things was meaningless. To this day it is the loss of personal effects that affect me. Every day I have a need or a memory of something irreplaceable I once saved but no longer have. Photos, baby paraphernalia, school reports, letters and cards, my library from university and hundreds of favourite LP records from years of collecting music.

Worse was to come. We were fully insured for contents. The insurance company refused to pay out. I learned later from a colleague in the insurance industry that the head of that company had a standard approach to all claims, to simply refuse all of them and wait to see who succeeded in challenging it, which I was unable to do. To this day I am reluctant to take out insurance, as critical as it is in my business world. I do but I hate it. It gets me all fired up.

CHAPTER 5

PIPE DREAMS

I built a practical robot suitcase for Samsonite into one of their Piggyback models. By now I was travelling overseas constantly and my favourite suitcase was the innovative Piggyback by Samsonite, the most recognisable brand back then; a medium-sized case with wheels when not all suitcases had wheels, and a broad wrap-around strap integrated into the pull-out handle so that it could hold a second, third or even more, suitcases, on its back, becoming its own trolley. I remember thinking, Why don't all suitcases have wheels?

It was a decade or so after I built the original Tasman Turtle robots, and I had already designed a number of robots for other companies, incorporating my technology into them under licence. It was normal for me to scan the trade magazines looking for ideas and companies to contact to generate more business. One day

I came across a fluff piece about motorised suitcases that made moving heavy luggage easier, so I decided to contact Samsonite.

Samsonite is a multinational, so finding who to deal with was not easy. They are now in Massachusetts, but back then they were based in Denver, Colorado, where the company had been born. I chased them down, sent an email that got the run around, took a long time, but ended up with their legal counsel, generating some interest. Most of my unsolicited contacts with larger companies ended up with the lawyers. I learned early about the risk of intellectual property (IP) theft, or conflict of interest, or simply scams, so now my unsolicited approaches start with a legal disclaimer, but still they all progressed to legal departments for decisions about whether to answer me or not. Samsonite was no exception. These days I use a short cut and send my letters straight to the lawyers. They like that.

A suitcase is a suitcase is a suitcase, so a company like Samsonite creates higher value through branding, and that means IP like design patents, logos and registered trademarks, and that meant lawyers, who at Samsonite came in the form of Bill. Tall, gangly, more Midwestern than Western, Bill turned out to be as unlawyerly-like as is possible; friendly, approachable, seemingly unguarded. We became friends and I discovered he was also a frustrated inventor. He had a clever travel-related product idea and in me he had found a person to turn it into a reality. So, eventually, as well as the suitcase robot we built a non-robotic tech device and had a global patent together.

I decided to fly to Denver, mile high city, the same elevation as the highest mountain in Tasmania, Mt Ossa. I was to be met at the airport by an officer from the company, at the then brand new, inconveniently massive airport outside Denver. I hate large airports, especially those in Australia, designed for everyone except the traveller by what seems to me to be greed merchants: those lengthy, unnecessary walks which force you past shops you don't need; no facility nearby to be picked up outside, so more long walks. It's exhausting after long flights, and unbearable for the elderly and infirm, who you see struggling with the unnecessarily poor designs.

One time, I was in a shop beside a canal in Amsterdam and a peculiarly strange man beside the man serving me, maybe a friend but clearly not part of the staff, and who had been completely mute the whole time I was in the shop searching, asked me where I was from when he heard my accent. I said Tasmania. He said 'Mt Ossa', then nothing further. I've thought about that and believe he was probably an autistic savant, someone with Rain Man-type knowledge and memory of facts, at the loss of other social or communication skills. He probably knew the tallest mountains everywhere. The Dutch discovered Tasmania, and the first mountains to be named, seen by Abel Tasman as he sailed down the turbulent west coast, are Heemskerck and Zeehaen – after his ships – although Anglicised now. But Mt Ossa is named after its Greek predecessor.

Making plans for my arrival in Denver, the Samsonite officer and I had never met each other, so jokingly I said, 'I will be recognisable because I am short, fat and ugly.'

When we met, the person said, 'You lied.'

'How?'

'You're not fat!'

When I met with them in their head office meeting room, all of them jovial and convivial as if it had been in their job selection criteria, I suggested to them a suitcase that could follow you around through airports, but they were not interested. Instead they said that the one thing they really needed was a suitcase that could get itself over kerbs, up stairs, even maybe into the trunk of a car. So, I said that of course I can do that. An executive at Samsonite had sprung his back hauling a heavy suitcase from the car to the departures entrance of an airport, which explained this request. It also smelled of lawyers to me again, controlling the risk of potential liability lawsuits.

The clincher was that the mechanism of such a robot suitcase had to be invisible inside the suitcase, take up almost no room, not weigh more than a couple pounds or so, and cost no more than $11 in mass production. 'I can do that.' My very simplest robots at that time in mass production cost around $100 to manufacture. Most others were state-of-the-art sophisticated machines and much more expensive.

But I did it. Back in Tasmania where our research headquarters were still based, we received from Denver a huge pallet of various

Samsonite suitcases to use for experimenting. Way too many. Some of those served us well as personal luggage for years to come. Until airlines became so weight conscious that Piggybacks were considered to be too heavy and phased out. We created a robot suitcase with motorised drive wheels, the motors hidden in the external cowling of the existing wheels; flexible gel cell batteries packed inside against the base of the suitcase like an inner skin; clutches on the drive mechanisms so that when the motor was not actuated the wheels could still turn; and a dead man's switch in the pressure sensitive handle, so that if you dropped the suitcase it would not careen down the runway like a pompous haute couture model. The wheel motors also drove a set of tracks on the front of the case, to clamber over step edges when you pulled the handle switch. All of the parts came to around $11.00 and occupied none of the internal space of the case.

This led a few months later to introducing the first prototype back at their head office in Denver, where it was filmed. Later, the second prototype was presented at their board meeting in the Flemish city of Oudenaarde in Belgium. Our meeting there, in a modern glass-windowed building, was in the tallest and only structure of its kind in this ancient town of dark-brown beer and medieval-looking, delft blue tapestries. At that second meeting Chairman Steve Green was impressed with the solution to the difficult specifications and said he had another little company in Chicago and arranged for me to go visit them.

'What for?' I asked and mysteriously he said, 'Just have a look.'

To my mind Steve Green was a kind man in the midst of ruthless corporate raiders. While raiders by definition tear companies apart to sell off bits for a profit, with no concern for anything else, Steve was a builder of companies. It was at the time of the Carl Icahn, Leon Black takeover battles, and the juggling of company brands like American Tourister, Astram and Samsonite. Like a foster kid, Samsonite was shuffled from company to company, investor to investor. Very incestuous, and my interesting introduction to the high-end private equity sector. Steve Green's other 'little company' was Culligan.

So, on a dull grey day I turned up at Culligan in Rosemont, Illinois, to an unassuming dull grey industrial building, home to the largest commercial and industrial water treatment equipment manufacturer in the USA, maybe the world. Somewhat confused and lost for an agenda, I met the CEO, who was, I presume, equally confused, but also somewhat infuriated that a board director had instructed him to host me. Culligan's large municipal water purification plants were like giant silos, or medium industrial installations – large welding cylinders that, for instance, supplied water to fast food outlets at truck stops along interstate highways. He asked me what I was there for and I told him the story. I felt he was more interested in my robots than his water. By default I received a guided tour of sales, production, R&D, administration, through this typical old American industrial manufacturer, and returned to his office. None of us knew what I was doing there, especially me.

Lost for something to say, I asked him what their most pressing problem was, right at that moment. He said that there was an emerging market for do-it-yourself, under-the-sink water filtration and purification systems. They had designed and released a couple of DIY test products, but they were fraught with leakage issues at the hose connections, and recalls and returns were a serious challenge. When I looked at their products, I laughed. He was not bemused. He asked what I was laughing at. I said it was obvious. They were miniature clones of their large industrial plants. Cylinders and couplings, connectors and chemicals, pipes and tubes like a row of vacuum flasks, all hooked up together like a patient in an ICU unit at hospital. I was told that no matter how they designed them, no matter how carefully the instructions were worded for installing them, the customers had difficulty and they leaked. Salespeople received complaints.

My method of solving problems is always to go back to basics, the further back and the more basic the better. Also, I believed firmly that too many parts in any product design means unnecessary complexity, and complexity means more things to go wrong. My mind wandered. On my rural property in Tasmania I had a lot of frustrating experiences with never-ending leaking hoses and pipes and connectors, and had thought about, but never pursued, a novel idea I had for leak-proof connectors. Back to reality in Chicago, I had an instant idea for Culligan which meant I would need to depart to do some shopping. I arranged to come back the next day.

Holed up at a nearby cheap motel, I drove to the nearest hardware store, bought a foot of 5-inch round timber, and asked for it to be cut into thin discs about 2 inches long. I ended up with short and fat pieces like largish ice hockey pucks. I bought some snap-lock garden hose connectors with those black rubber O-rings which you push in or pull out with no threads or metal clips, some sandpaper and a spray can of matt black paint. Before leaving Culligan, I had collected a couple of bits and pieces from the R&D department floor, including some branding labels.

In my motel room I smoothed the wooden discs before spraying them with quick drying paint. Then I cut and glued the rubber hose nipples and sockets to opposite sides of the wooden discs. On their side, they looked a bit like those training wheels with a handle on each side to hang onto and roll out across the floor to strengthen your back. I had forgotten to buy a hacksaw blade, so I blunted a kitchen knife when cutting up the hose fittings, Finally, I pasted some Culligan logos on the side of the discs. Discs could be snapped together snuggly and flush on top of each other, connected automatically by the nipples, which become invisible between discs in the stack. There was still one hose to the top receiving water in, and one hose at the bottom feeding filtered water out. Or the other way around, not sure now, but that was an improvement on the previous configuration. The different discs, theoretically, were to become replaceable consumables, added revenue streams for Culligan, with different contents: charcoal, sand, mesh, chlorine; whatever was needed in the series of filtration processes to remove sulphur, particles, heavy metals, bacteria. Stack as many as needed. My sample had only two or three discs to illustrate the point. The design called for a U-clamp to go around the discs to hold them together, but I could not find a suitable one at the hardware store. It looked great, had cost a few dollars and only took a few hours to fabricate a prototype. It took longer to remove the paint from my hands and the smell of it from the room.

Next day at the CEO's office, I said I had a solution, a novel design, for their DIY sink filter problem, and brought it out of my Samsonite briefcase, hidden in a brown paper bag as if it were something smutty. Talk about professionalism. They asked what it was, and I said they could only see it if they paid for it in advance. Talk about chutzpa. I said that they would own the prototype and any IP rights, but that they had to trust me and pay for it first. This was around 1990, and I nervously asked for $35k. Maybe it was the fact that Steve Green had recommended me, or something about my naked Australian bravado, but they wrote a cheque out then and there. I nervously took the wooden sample out – luckily now dry and not tacky from the paint job – and placed it on the CEO's office desk in front of him. The CEO was ecstatic. I now know I undervalued it. I also know now they could have easily torn up the cheque regardless of the arrangement if they had been disappointed, but I never thought about that at the time. The R&D manager who was present, and who had developed all the leaking models, was understandably unimpressed, saying anything and everything he could think of to detract from the originality of my impromptu design. But there was no doubt it was a winner.

With a smug grin on my face, a cheque in my pocket, an expectation of more to come, and Billy Ray Cyrus's 'Achy Breaky Heart' just released and very loud on the car radio, I drove back across Chicago, my head as high in the clouds as the vertiginous edifices that dominated this home of the world's first skyscraper. Feeling that this simple Vandemonian was holding his own in the most aggressive market driven economy in the world.

New York is thought of as the home of skyscrapers. But the long-gone Home Insurance Building constructed in 1885 in Chicago, twelve storeys suspended on an iron framework, was the world's first true skyscraper, predating the beautiful and extant Flat Iron Building, the first skyscraper in New York. Traditional building technology, bricks on bricks or even concrete pillars, gave way under their weight after about eight storeys, and it was the idea of using a steel framework, a metal skeleton, that allowed the weight to be supported by the towers of today. Those edifices sky scraping, scratching the heavens, almost a kilometre straight up, above the clouds at times. Before this architectural technology, at about eight storeys even churches needed buttressing to stop the walls from collapsing out. And even early pyramids, built at too steep an angle, collapsed under their own weight, like a pile of heaped sand settling to the right shape.

'Achy Breaky Heart' was released in the USA on 23 March 1992, so that dates this memory. Nothing makes me feel older, quicker, than to realise modern music is just noise to my ears and my favourite songs are just noise to others.

CHAPTER 6

OPTIMUM PERFORMANCE

Everyone has their group of closest friends. There is evidence that the number of friends you have is controlled by a gene. Not a gene for one or two or three, but a gene for a few or a lot. I have a gene for a lot. That makes sense. There are gregarious people and there are those who are inhibited, those who like to party and those who are comfortable in their own space. Like many people I make friends relatively easily, but the ones I call true friends, that I'd do almost anything for, number fewer than the fingers on my hands. English does not seem to have a word for those deep friendships versus the general run-of-the-mill type.

Mine include more than a proportionate number of older people: a teacher, a professor, a boss, a business colleague. Then there's a workmate, a schoolmate. The older ones are not quite a generation different in age from me, so they are a hybrid of father-figure, surrogate brother, friend. Some have died, but I sincerely hope that they all received as much from having me as a younger friend as I did from them.

The teacher was the German Eddie Sauer, who I helped build car engines that ran on water and helped me to think like a European. The professor is Jon Jarvik in Pittsburgh, who like a true scientist, challenges everything I say, keeping me honest. The boss was Max Rowbottom, who taught me to be an operating theatre technician and who helped me make precision machine parts for some of my early robot prototypes, and who committed suicide several years later; not the first person to do so in my life, or the last. The business colleague was Zack Rosenfield, a quiet, unassuming architect in New York and uncle to a girlfriend. The workmate is Alex Vail, the coolest character I know, the most assured person I know; a Scot who arrived in Tasmania when he was about 16 years old. We played in bands together, rented flats together, worked together, travelled together, dated sisters at one time. The schoolmate is Ron Davies, actually a university mate who was in my first year at medical school and taught me to scuba dive, and who, with his family, shared our camping escapades. These are all males, but there are females too. Robin, another professor who named her boy after me; Dabni, who taught me to be professional in all I do; Faye, who taught me not to become complacent with the Jewish holocaust; Sabine, who supported me then and supports me to this day; Ruth, who dedicated one of her scientific books to me; Mim, the love of my life but with whom it never worked out. A rarity is a new best friend from a handful of years back, Judi, who is as different from me as possible, yet the rapport is as strong as possible. That's because the basics are there: honesty, generosity, caring, true-to-self. There are a small number of others, so this is representative.

Zack, handsome, medium height, slim without effort, whose hair never really ever went grey, came into my life later, but was to become my very favourite person of all. He was one of those rare individuals who people listened to, not because he was loud, but because he whispered.

My story with Zack starts earlier. In 1983 I moved to Texas as head of R&D at the Commodore Computers research facility in Dallas, Texas. One of my colleagues there was Richard Wiggins, who previously had been at Texas Instruments, a company of firsts located also in Dallas, across town. Texas Instruments created the first commercial transistor, first transistor radio, invented the integrated circuit, and released the world's first hand held electronic calculator. Richard, who looked a lot like handsome actor Richard Chamberlain in *Dr. Kildare*, had been part of the team that developed their revolutionary Speak-and-Spell, the colourful talking calculator for students. Prior to Dallas I had met captains of the technology industry like Jack Tramiel, the founder of Commodore, but Richard was the first top of their field international technologist I had come across, and like most people in such positions he was kind, made me feel welcome, treated me like an equal and was genuinely interested in who I was and what I was bringing to Commodore. This unrestrained egalitarianism was a refreshingly new experience for me. I was to discover as the years went by that it was very rare that I worked with anyone who was insecure enough to intimidate or look down on me. For a young upstart who had always been trying to prove themselves in a small pond, to be accepted in a big pond was confidence building.

While at Commodore in Dallas I invited the Honourable Robin Gray MP, the Premier of my home state, to visit as part of a North American trade mission I convened. Highly controversial, Robin Gray was nevertheless a state development proponent and my robotics technology startup in 1979 was probably Tasmania's first true entrepreneurial, advanced technology venture. New ground, so the government was very interested. Governments are always very interested in riding on the back of someone else's success. Acting like it could not have happened without them. With me, though, they were on a learning curve. I was only one step ahead, but they did not know that. At one time, when presenting a commission payment formula for technology transfer, one of the government officials asked me if this was an industry standard. There had been no experience with high tech in Tasmania and many politicians and bureaucrats and industry leaders frequently visited and consulted with me. I'd teach them. I felt like Mrs Hamilton, my schoolteacher.

The Premier and his minders stayed in Dallas a few days, and were guided through the activities at Commodore, seeing first-hand the complex technology of designing and building computers, speech processing technology and silicon chip making. The entourage was similarly fascinated by the infrastructure in Dallas, and with sights like Dealey Plaza, the assassination site of President John F Kennedy. Premier Gray was fascinated with the glass-faced corporate buildings in Dallas, suggesting power and wealth. Some of those ideas appeared in Hobart soon afterwards, like the Grand Chancellor Hotel. From there I took the Premier and his entourage to meetings with the company founder, Jack Tramiel, in West Chester, PA.

Commodore had embarked on a hobby robot development project, and discovered along the way, like everyone else, that this was no easy task. Their project had spent a fortune and was floundering. Jack Tramiel and I had met in Hong Kong, where I was building robots for global electronics giant Radio Shack and for Tomy Corporation, one of Japan's largest and most innovative toy companies. He was impressed with these projects, so with offers of money, he attracted me to Dallas.

Despite being offered what I subsequently learned was the highest salary at the R&D centre in Dallas, before accepting the position, I remember calling Jack and saying, 'I also need a $50,000 "start fee".' This was a huge amount for 1983, when my salary in Tasmania had been around $16,000. I had just read about such contractual perquisites and felt it would be worth a try, which it was. I was becoming Americanised.

All of my robots until then, some of them quite large and expensive multifunctional hobby robots, nevertheless displayed simple mobility, dead reckoning, touch and move reflexes and such, but none of them had any true autonomous navigation capability. This was something Commodore wanted in their robot. No robot in the world at this time had this technology in any practical sense.

Already I had started realising that the approaches to robotic technology had taken two turns. One was the way everyone was doing it: physical sciences, hard engineering, intense mathematics, extensive software coding, with additional software routines added one-by-one to deal with limitations or exceptions detected in testing. Unwieldy. The other was the approach I had taken: organic, intuitive, analysing the way nature had already done it, which was influenced by my previous experiences in medicine, biology, psychology and surgery. The first way I called brute force. The second way I called elegant, not reinventing the wheel. Commodore had been using brute force, which is expensive.

The industrial background of the engineering team I joined was visibly evident in the out-of-place and impractical microwave dish, with its yawning aperture, almost as large as the robot itself, that they had designed and fabricated and mounted on the top of the prototype. It was the transponder (transmitter and receiver) to map the environment in a radar-like fashion. They were proud of it, but to me it was like those penny arcade stalls where you put a ball into the gaping mouth of rotating clown heads. It was like that disgusting image that you see floating around on the internet of a large ear on the back of a mouse. The dish was derived from the technology of telephony microwave dishes on repeater towers, using ultrasound instead of microwaves, using echolocation to try to map the surroundings, admittedly with reasonable success, but totally impractical.

For the transponder to be practical it was essential that it had to be small, cheap and functional. Commodore's team had made no headway with that challenge. Then there was the mapping problem. Getting a bunch of echo points from such a sonar transponder was one thing. To build reliable maps like floor plans from them and then use those to navigate autonomously was another. Solving this problem, to develop this capability, was a task they had contracted to Jim Crowley at the Robotics Institute at Carnegie Mellon University in Pittsburgh, CMU. On my visits to CMU I discovered that their approach to the project was also expensive, progressing slowly and not producing useful results. I could see the prescience of Jack Tramiel choosing me for this job, and the instinctiveness that he had used to create the world's largest PC company. He knew his teams were not getting anywhere, but not really being a technologist he did not know why. I think he felt that I might be the alternative approach they needed. It's an internal 'feeling' thing, he told me years later.

In my view, the brute force approach paid little attention to world context; that awareness that comes from knowledge of the world and its multidimensional connections, causes and effects, consequences, influences … some of which would be learned, some of which might be innate. Humans are born with an innate fear of snakes, for instance, and babies are born recognising a smile. That became my gambit. I recognised a fatal flaw in the CMU mapping algorithm which led to maps that 'wandered'. The maps were like a floor plan. Instead of being robust though, some representations in the maps, fixed room walls for instance, moved around as if they floated, or shifted as if there was an earthquake happening. I was soon to solve these issues back in Tasmania in a robot called Blinker, named after a fictional wombat from a sad children's story we all grew up with.

But right now Commodore needed a product. Anything to recover the investment to date. All of the experience from my earlier robots now went into the prototype back at Commodore which we named Chester, in honour of Commodore's head office in West Chester County, Pennsylvania. Chester was completed and released to the world at the Consumer Electronics Trade Show in Colorado in 1984, serving coffee, with the clear, hemispherical, Tasman Turtle trademark dome evident as a legacy on its head, but without the autonomous mapping, or thankfully, the giant ear!

Years earlier, when Commodore was a simpler company manufacturing mechanical accounting machines and was experiencing financial troubles, a silent investor was brought in. At the time of Chester a dispute between Jack Tramiel and that silent but largest shareholder of Commodore, Irving Gould, in

Toronto, led to the departure of Jack, the collapse of Commodore and the end to their autonomous robot aspirations. I returned to Tasmania and continued to finish the self-navigating robotics technology which then became the foundation of all my subsequent industrial and commercial robots.

My ex-boss and friend Max Rowbottom, a skilled machinist, made the tiny rotating ultrasound mechanism for me that superseded the unwieldy Commodore dish. Instead of a massive gaping cavity, mine was a small disc the size of an actual ear.

Jack went on to buy Atari, which had been founded by Nolan Bushnell, and which had also fallen into decline after it was sold. In 1987 I worked with Nolan at another of his many companies, this one called Axlon in Sunnyvale, California, the heart of Silicon Valley, but that is another story.

At CMU during the visits to Jim Crowley to be updated on his progress with the Commodore project, I met one of my heroes, one of the few genuine pioneers of robotics, Hans Moravec, who built the famous Stanford Cart. This robot and perhaps the Johns Hopkins Beast and Shakey were the only robots in the world in the late 1960s and early 1970s. They were exciting, visionary, one-off research vehicles, not commercial products.

The scientific write-ups about these pioneering robots reads well, but the reality was different, again showing just how difficult robotics was, even for these technology giants. The Stanford Cart was built on a shoe-string budget to use vision to see and plan a path around objects to reach a goal position. It performed a complex scene analysis and navigation task, but only a step at a time, taking hours to traverse a few yards, as the huge software programs had to be loaded and unloaded from the limited memory space. It would take a look with its single camera, which because of the lack of funds had to slide along a rack to a second position to get a second image to regenerate the stereo information. The stereo vision then created the depth map, the distance to objects the robot had to circumnavigate. Then the program was changed to one that could plot a path, then changed again to drive the wheels.

The Johns Hopkins Beast roamed around corridors randomly until its batteries needed recharging, when it searched for the nearest special electric charging outlets, just like current electronic vehicles have to do. But that was its life, and at times it would still fail in this simplified environment that had even been modified to benefit it.

Shakey was so-named because its motion control was so unstable it shook as it moved. But there was nothing like these robots before and they were fascinating.

Because of these CMU connections, in 1989 I was invited to a visiting professorship position at Carnegie Mellon University in Pittsburgh by Hans, who said to the university that he rarely had an interest in working with anyone, but was making an exception for me. With no advanced degree, normally a prerequisite, they recognised my career PhD equivalency and appointed me formally as Invited Visiting Scientist.

More of the most amazing industry and academic connections opened up. Universities like MIT, Harvard and CMU have massive promotion and publicity machines, dwarfing anything I had seen in Australia. One day I had a call from *House and Garden* Magazine, about Isaac Asimov, the profuse science fiction author. Isaac had documented his famous Three Laws of Robotics, and he had written many science fiction robot stories including *I, Robot* around 1940, not very long after Czech playwright Karel Capek in 1920 first used the Germanic word 'robota', meaning 'slave labour', in the sense we use it today. Asimov had been asked to write an article for the magazine about robotic homes of the future, but when I talked with him it emerged that he was very ill and I was asked to write the article in his place. Unfortunately, Isaac passed away shortly afterwards and nothing came of the article. Isaac was a thought leader. *I, Robot's* protagonist was Dr Susan Calvin, decades before gender equality. It is also the inspiration for the name of the robot vacuum cleaner company iRobot.

Another one was the invitation by the R&D division of General Motors in Detroit to come and present a workshop on driverless cars. These were heady days.

Ann Hornaday contacted me one day from New York. Flaming red hair, dimples to die for and as smart as a rubber band snap in your fingers. Ann was a freelance writer and journalist doing a piece for *Omni* Magazine on robotics. *Omni* was a fascinating monthly magazine that I subscribed to back then. Eponymously,

Omni had articles about everything, from science, to science fiction to sacrilege. We had met at a dinner party in New York earlier, and we met again for lunch in the city. She paid, courtesy of *Omni*. She explained they wanted a simple one-page construction project for the reader to build their own robot. The trick was that these would be non-technical readers, so no soldering or knowledge of electronics could be used. Sounded impossible, but I said, 'I can do that.'

Sometime earlier I had read a piece in a staid and dry science magazine about artifacts discovered at an ancient archaeological site, which when analysed showed that the prehistoric culture had made simple mechanical analogue computers. Images of it were reminiscent to me of the amazing mechanical analogue computer built from Meccano set pieces to calculate the trajectories of Barnes Wallis's bouncing bombs, and which I had seen in a museum in Auckland, New Zealand, being a life-long advocate of Meccano. Plus there was of course the Antikythera mechanism of Ancient Greece. Not to be left out, for fun I designed a system of logical gates inserted into matchboxes, using lollypop sticks, rubber bands, drawing pins and duplicating the functions of Boolean logic: AND, NAND, OR, NOR, NOT, EXCOR and all the other binary functions that normally come in little black TTL and CMOS silicon chips. It was easy to see why *Omni* did not want to get too technical. My little mechanical black boxes worked; for example, pushing two lollypop sticks into the matchbox was necessary before a stick protruded from the other side, popping out in an AND function. With my successful robotics articles in *Electronics Today International* I thought they'd love that stuff, so I sent it off to them for publication, but they rejected it. To my additional embarrassment, a few days later it turned out that the prehistorical ones I had read about were an April Fools' Day joke in the normally serious magazine, which had completely sucked me in. But I still liked my little tactile mechanical logic boxes and was desperate to do something with them.

One of the demonstration programs provided with the Tasman Turtle was a replication of classical conditioning, also called Pavlovian conditioning, as described by Ivan Pavlov, whose original book I had read. The favourite Australian desert is a meringue-like thing called a pavlova and I am conditioned to them. It is important to me to read original classic works, like Dr Lawrence Peter's *The Peter Principle* or Malcolm Gladwell's *The Tipping Point*, in order to avoid miscomprehensions or misquotes. It's another form of my going back to basics, my modus operandi. Pavlovian conditioning contains elements of learning and memory from sensory interactions so is close to my work in smart robotics.

The Turtle robot had a bumper bar; a sensor ring all the way around its circumference, and pushing it operated touch sensor switches that indicated the location of the push. In this demonstration the robot was programmed to roam around randomly and if it bumped into an object, it would sense the object with the front touch sensor, stop, turn around and drive off again in another random direction. An unconditioned reflex.

Exactly like the current toylike vacuum cleaner robots do. If the back touch sensor was touched, it had no effect on the behaviour of the robot. The back touch sensor could be touched forever to no effect. A neutral stimulus. However, if the back sensor was tapped just before the robot bumped into something – the timing was critical – like a warning tap on the back, initially the robot's behaviour was unchanged, but eventually it seemed to learn about the warning, and it would stop before it bumped into anything and then turn around immediately to go somewhere else. It had been conditioned with a new reflex to avoid a collision.

The program to do this was also easy to duplicate in simple binary logic, and a few of my mechanical gates could do it. The next meeting with Ann Hornaday was in an office at the striking address of 2000 Broadway. While I was presenting the idea to her in a spare office, there were semi-naked women roaming around outside the windows of the office. I did not know until then that *Omni* was published by Bob Guccione, of *Penthouse* fame, and photo shoots were in progress on the same floor as we were meeting. These days Ann is a prominent movie critic for the *Washington Post*, but that day she inadvertently treated me to one of the perks of working in New York.

One day as I was walking down the street in Midtown, Manhattan, the business heart of New York, I saw Ann up ahead, so I raced up to tap her on the shoulder. When the person turned around it was not Ann, and embarrassed, I apologised and said, 'Sorry. I thought you were someone else.'

'I am someone else,' was the reply.

Like *ETI* before it, *Omni* rejected my suggestion. Later when I bought a copy of *Omni* to see what they had done instead, it was a radio-controlled car, with modified electronic circuit boards that needed soldering in their construction, and which roamed around as an almost trivial remote-controlled vehicle. As boring and as unimaginative as one could get, and unworthy of *Omni*. The project had been done by MIT, who had found out about the project and convinced *Omni* to take them on, with their superior name, instead of me. MIT's promotion machine was greater than CMU's or mine. The article was unworthy, in my opinion, of MIT too.

At around this time on a memorable day at CMU there was an evening function when someone I had never met came up to me, introduced herself as Amy and chatted. She was beautiful, smart, and I was immediately smitten. I asked if I could give her my business card, hoping of course that she might call me. As I passed it to her she dropped it, accidentally, and I remember jesting that she is supposed to wait until I am not looking before throwing it away. We laughed and the next day she called me. We dated for a long time in Pittsburgh, but it did not work out. Her uncle was Zack Rosenfield, to become my best friend in the world.

Zack, who I did not yet know well then, although I had met him a couple of times with Amy and who allowed us to holiday at his beach house in Deer Isle, Maine, called me from New York one day asking if I could help him with a problem. In his wooden home on a large block in a private cul-de-sac in the Bronx in NY, Zack explained that he had recently retired from his private architectural firm in the heart of New York City. He'd had a couple of multimillion dollar retirement funds, but both funds had been embezzled, in different ways by different people, and at 70 years of age he had no money. The culprits had been caught and one was already in gaol, but the money was gone, so it was little consolation. By then I had left the robotics industry that I had pioneered, and was already busy turning around other companies, which Zack knew about. So he asked if I could help him.

Amy and I had split, acrimoniously unfortunately, and my involvement with her uncle made us both enemies with her. To top it off, Zack said he had just been diagnosed with prostate cancer. Being connected as he was with the medical fraternity, he was receiving leading-edge treatment with new drugs and he explained that everything was under control, but what a sad sequence of events. Helping him through this time was a no-brainer.

I researched Zack to determine what could be done, and discovered his fascinating background. He had studied physics at Columbia University in NY, he was a NY native, but did a second degree at MIT's Space Planning and Organisation Research Group, known as SPORG, in Cambridge, MA. It was all about the 'built environment'. He was well known in the industry for inventing several innovations for hospitals.

Zack's was a family of architects. His father was one of the NY city planners and responsible for the design of many high-rise hospitals and medical facilities. This was the legacy that Zack grew up with, and in turn he had been written up as the creator of innovations such as what are known as clusters wards, where a nursing station in a hospital is central and in sight of all the wards, for obvious safety and risk control. Also, he advocated subdued, hidden lighting, wide corridors, more green plants and other inexpensive aspects to creating healthy and happy facilities, encouraging rapid recovery and productive work. I could use all of that to help his recovery from what he described colourfully as financial meltdown.

Zack was prevented by the noncompete provisions in the contract around the sale of his business from returning to architecture. Instead he had attempted to work as a consultant to aged care homes, but after a couple of years and dozens of meetings, had not secured a single paying contract. When I looked at his presentation material, none of that fascinating and invaluable background was included, so self-effacing he was, and so there was no hook to his sales process.

Excited by this marketing ammunition, I immediately revamped his printed material, revitalised his PowerPoint presentation to promote how skilled he was, created a website, and rehearsed with him presenting it. We then started a round of new meetings at various aged care facilities in the New York, New Jersey, Connecticut tristate area. We started getting interest.

The next thing we did was to create a monthly newsletter, with stimulating sections relevant to the aged care industry, a bit of self-promotion material that I had to write, book reviews, the odd technical piece, and a few light-hearted fun pieces. The only thing missing was Sudoku. This not only went global, in hard copy as well as electronic, but we received calls from people we did not know requesting copies or subscriptions. Then we discovered that relevant academic research was being conducted at Rochester University in upstate New York, so up there we went, formed an alliance, and began incorporating that connection into our presentation material.

It was on one of those Rochester trips that I combined my proprietary innovations on Standard Structure for companies into the budgeting for nursing homes.

To my surprise, when I first started turning around stressed companies, and studied Organisational Structure, I discovered in my analyses that corporate structure did not vary from company to company as the literature dictated. Being who I am, always back to basics, I satisfied myself as to the reason this might be the case. It was one of the most satisfying discoveries of my corporate life. On the surface structures looked different, had different names, but a deeper study, along with an awareness of the evolution of what a company is and what it is for, confirmed otherwise. I formulated what I then called Standard Structure as the fundamental internal structure of all successful companies. I went as far as to say that if you have Standard Structure it is no guarantee of corporate success, but if you don't have it, it guarantees failure.

The idea to use this in financial budgeting of aged care facilities emerged as we were brainstorming one morning while having breakfast at a diner next to our motel ahead of a meeting with scientists at Rochester University. It was drafted literally on the back of a napkin, but in fully developed form was called The Optimal Performance System, or TOPS, and published in a peer-reviewed academic journal. Nursing homes are all different from each other, just like companies are different, with varying budgets, resident characteristics, socioeconomic profiles, boards of directors, and so on, but they all had one purpose, which was to care for their residents. So, like companies, a standard structure could be used to determine the best use of funds to provide the highest quality of life for residents under their differing circumstances.

Zack became my closest friend. This work was carried out over a 15-year period and I visited New York several times a year, usually as a house guest, and we worked together as a team. It was more than a job, it became part of our lives, and the shared companionship was as much the reward as the recovery of Zack's finances. It was also the first time, but not the last time, that someone told me afterwards that I saved their life, such can be the depth of despair from financial failure.

I never knew a calmer, more decent person than Zack. He was a humanist, which I had not known; he introduced me to the Ethical Culture Society in New York with its famous Fieldston School; to any number of prominent people, famous even, who were his neighbours in and around Riverdale, his neighbourhood in the Bronx; to business opportunities for my company; to see expat cricket matches at nearby Van Cortlandt Park. From his house it was possible to look across the Hudson to the impressive New Jersey Palisades, swim in their local pool, or walk to famous Wave Hill horticultural park, passing the imposing, white JFK house. Genuine, unblemished integrity, clever as it is possible to be, surprisingly conversant with the classics (unusual for a scientific mind), able to describe the characters or plots of Shakespeare, or who is who in the worlds of the Norse or Greek gods, and with the strongest gift for spelling I have ever come across. It confuses me why he took me under his wing so totally.

One sad day while I was back in Australia, I received a telephone call from him from Maine where he was now living, saying he had been diagnosed with leukemia, a version that was the result of the prostate cancer medication that had kept that disease under control, and letting me know that he had only weeks to live. He sounded strong and healthy, and the news was unbelievable, unbearable. He agreed. He said he felt no different, and it was surreal, to use his word, to know he would die very soon. Zack passed away a few days later and I never got to see him again nor attend his funeral. A mutual friend, Leslie Goldberg, read a short eulogy for me at his Ethical Culture Society memorial. My best friend was gone and has not been replaced.

Many afternoons, Zack and his wife Marydel would make a soup. They did many things together as a duo and this culinary exercise was a fixture in their lives. The room would swell with the sweet aroma of cooking vegetables. I also learned from them to make a soup of spinach and leftover cold rice, and was surprised to later learn it has a Greek name: spanakoryzo. What they cooked, though, was always the same soup – onions, carrots, beans, spices, no meat – yet every time they embarked on it in the kitchen together, they pulled out the recipe first and checked it. Then they would start to argue about the ingredients and the process. It was quite curious to observe. Surely, they knew the recipe by heart. I know the recipe by heart! Zack was always calm and passive, so this kitchen behaviour was out of character and one day I broached the issue with him. Marydel was more volatile, so nothing strange there, but Zack was the one displaying altered behaviour. This particular day before they started their little culinary project I suggested to Zack that he eat a slice of bread and butter, which he resisted. He rarely snacked between meals. But I insisted and he did. I had not explained why. That time as they concocted their broth there was no argument. So unconscious was this behaviour that they did not even realise the difference until I pointed it out. I said to Zack that he was hypoglycaemic, and that by late afternoon when he was starting to prepare for their evening meal, his blood sugar was low, which made him cranky. I want Zack to be perfect, but this shows he was human after all.

CHAPTER 7

AL FRESCO

We were slowly sipping thick, invigorating coffees at a large, bustling street cafe on the Avenue des Champs-Elysée, watching people streaming by in equally thick crowds on this late sunny Parisienne morning. I love people-watching from street cafes around the world, be it in Paris, Melbourne, Hong Kong or New York.

Years of living in New York, back in the day when there were newspapers, reading the *NY Times* on Sunday cover to cover, the book reviews, the crossword, used to be my favourite weekend pastime, but newspapers have gone and cafes have not.

Nothing is more enjoyable to me than relaxed brunches, with lively conversations and observing the eclectic mix of humans that is the flavour of any city like New York. Brunches can be simple diner food or the best this most cosmopolitan of cities can offer.

Not that I am an epicure, but I like certain foods and eat slowly, savouring the moments. At home I am in control of what I eat, no red meat, no processed foods – but when I am out I eat what everyone else eats, and love it. When I grow up I am going to open a bagel shop.

Even al fresco at home, on my remote rural property in Tasmania, tucked into its private valley, on the wide deck which I built (not as good as my dad's would have been), I love a complete brunch of too much food, too many competing aromas, too many choices, and a crossword or book. Or a deep, or not deep, topical discussion with visitors and houseguests. Or alone watching birds behave and animals misbehave and plants grow, not a care in the world nor a looming deadline, and soft music sinewing through the house out to the deck.

But Paris was different. Hectic. New York, for all its traffic and population, is casual. Paris is racing against time. Formal. The Champs-Elysée, or Elysium Fields in English, far from the setting of its ancient Greek word origins, unless Heaven for which Elysium is a prototype, is crowded and busy. On this day in Paris I began to notice that passers-by were stopping briefly to converse with a well-dressed man at a table near the footpath away from us. My sister Dianna and employee John were with me, and we were each noticing the same thing. Eventually the curiosity was too much for us and we called the waiter over to ask who that man was, must be someone special. The waiter was perplexed, he could not see anyone important where we were pointing. In stilted French we explained that he must be important because of what we had been seeing, but still the waiter could not identify anyone special.

Pointing almost rudely directly to the table, we explained that people were stopping to say hello and we thought he must be someone important. Eventually the waited recognised who we were pointing to and laughed.

He said, 'Oh no, he is just the restaurant owner, he is not important.'

We were surprised and asked why everyone was stopping to talk. The waiter laughed again and said we were right next door to the congress building and a TV studio and everyone who stopped was a celebrity, a movie star, politician or something. They were all very important people. We did not recognise any of them, of course. The waiter explained that they all had their coffees at this cafe, so were well known to the owner. The waiter was still chuckling at the strange Australians as he walked away. At least it's refreshing to know that others like al fresco.

Around the same time I was to fly to New York from Charles de Gaulle Airport. Sitting on a side seat near my economy departure gate and near some coffee shops, I was wasting time twiddling my thumbs like my Nan McKenzie used to do, and Uncle Gordon used to do, when an entourage of beefy men and a beautiful woman passed by and overtook some tables at the cafe near me. Somehow I managed to ask the Frenchman beside me if that was Vanessa Paradis. I had recently watched her in a movie on TV, typically French with a bit of nudity, and she also had a hit song on the radio, so I was aware of her, but I'd not heard of her before coming to France.

The man beside me was in wonder that I was talking about this celebrity and whispered, 'Oh, she is very famous.' Before he could do anything and before I could think how silly it was, I got up and walked over to their tables. They were still settling in. It was such a surprise and sudden move and so inappropriate, because everyone had seen her and been watching but politely keeping their distance, that her minders could not react in time as I started speaking with her. I got her attention and asked if she spoke English to which she smiled and replied she did.

I then said, 'I came all the way from Australia just to see you.'

Wide-eyed she said to me, 'No?'

And I said, 'No!' Then we all laughed for real.

The minders were still shuffling, realising their protection was too late, probably never expecting anyone to approach Vanessa. She was gracious and lovely and Chanel Coco was redolent. The only thing I had on me was a business card, so I secured an autograph from her and immediately bought her latest CD. I have them together in my collection to this day.

Later she married Johnny Depp, at least until 2012.

In Memphis, after a long overnight drive from Dallas, we stopped at a conveniently open early morning diner. Typical American diner like a railway carriage. Some are actually transformed railway carriages. No other customers were there and I asked for some breakfast eggs. The Memphis accent is incomprehensible, and so probably is the Australian English dialect. The brogues, theirs and mine, over the short order counter, was indecipherable and it took a long while to be understood. Eventually, though, beautiful, perfect eggs, perhaps the first time I used expressions like 'over-easy' or 'sunny-side-up', came along and we sat down to eat. The next patron came in, a black person, selected an old blues song on the jukebox, looked at us and had no problems ordering what he wanted, whatever that was; there was no way we could understand him. Then more people came in, all black, the server was black, all staring at us. Clearly we had wandered into a diner that white people never went to, and it was probably talked about for some time afterwards, and I am sure the foreign accents explained everything. We did not linger, despite the great music. Eggs 'al fasto' was what we achieved.

My friend Gordon Bellamy, an animation artist with his own way with words, used to say, 'If we had some ham, we could have some ham 'n eggs, if we had some eggs.' Words were different for Gordon. We were munching into spicy buffalo wings with celery and blue cheese dressing on the side when he said M.M.M. I asked what M.M.M meant and he explained: 'It spells mmm!' Then he asked me to pass the Whatsthishere Sauce. Lea and Perrins. He'd have loved it when years late I visited the famous Anchor Bar in the city of Buffalo, upstate New York, the birth place of buffalo wings, a recipe devised as a way to make use of usually wasted chicken wings. The restaurant ceiling was covered in heavy motorbikes, walls lined with number plates, the tables and bar covered in splashed red sauce and a real Wooden Indian monitoring the front door, just like Kaw Liga looking for his Indian Maid.

Gordon's dog Vincent used to always bark loudly when anyone came to their apartment door. Gordon tried to train him by putting him in the bathroom when he barked. After many trials, when someone came to the door, Vincent would still rush to the door and bark vociferously, but then put himself away in the bathroom. I often walked Vincent around lower NY, and after he was attacked by a Pitbull, I appeared on Judge Marilyn Milian's pseudo court room TV program 'The People's Court.'

On departure from the breakfast diner back in Memphis we had to do a couple of things. Get fuel for the car and find a cheap motel to sleep – we had driven all night. In those days I never booked ahead. Just drove along and pulled into any half-decent looking motel and took a room, $20 usually. First, though, we filled up at a self-serve and while inside to pay, I noticed the cashier was a tall, slim, beautiful, young black woman, with an out-of-character and patently painful limp.

I asked what had happened and she said, 'I got beat.'

Not understanding I asked, 'How?' perhaps thinking of domestic violence.

She pronounced it better and said, 'I got bit.'

I said 'Bit, like bitten?'

'Yes.'

I asked how and she replied, 'Ah got beat bah a daag,' as she limped back behind the counter to take my money.

The motels were a little more expensive here in Memphis closer to Graceland than earlier on the interstate, and we drove around quite a lot to find a cheaper one. Eventually we found one that was something like $15.50, a great price. So, I pulled up and went to the reception. Some people were already waiting, but we were served first nevertheless.

I said, 'I wanted a room please.'

She smiled and said, '$35.00.'

I asked, 'What about the $15.50?'

To which she looked sheepishly at me with a hint of a wink and a nudge, 'That is the hourly rate.'

I looked at the two or three women sitting there and said back to the receptionist, 'So this is *that* type of motel, is it?'

'Yes,' with unembarrassed grins. I think they had been looking just as much at me as I at them. My naiveté sometimes astounds me.

Another time back in New York, I arranged to meet with Robert Jarvik at a coffee shop near the Lincoln Centre. I was treating, as he was busy and going out of his way to meet me. One of my very best friends was Professor Jon Jarvik in Pittsburgh, a geneticist at Carnegie Mellon University, and it was a delight to learn that his brother was Robert Jarvik, the famous inventor of the Jarvik 7 artificial heart, by then up to the Jarvik 2000. The Jarvik 7 in 1982 was transplanted into Barney Clark, keeping him alive a further 112 days, not bad, but with even greater and greater success with subsequent patients.

One of my first jobs out of high school and before college and university was with Professor Robert Mitchell at the Department of Surgery at the University of Tasmania Medical School. Mitchell pioneered kidney transplants in Australia, and as a junior lab technician, I became an operating theatre technician, animal house technician, and involved in everything to do with surgery research, transplant technology, students and teaching, in this tiny department where any skill or talent you had could be deployed. Wish I still had that dream job. Under my boss Max Rowbottom (the first of several close people to commit suicide as it turned out later, including my dentist, my accountant, my sister, a girlfriend's sister, a landlady's husband, a neighbour), we designed and built an artificial heart-lung machine, a so-called cardiopulmonary bypass apparatus, plus some special operating devices. We had a dedicated tool shop, with everything except a milling machine, as part of our department. So, with this background, it would have been unbearable for me to know Jon Jarvik and not meet his brother. We had both built artificial hearts: his, a beautiful, shiny little inert metal implant the size of a thumb; mine, a machine the size of a fridge with pulsatile pumps, oxygen cylinders, filters, plastic oxygenation bags, and tubes in and out of the patient.

I learned that Robert had trouble developing the motor drive electronics for the electric pump engine in his miniature turbine Jarvik 2000 heart. So, as usual I said, 'I can do that.'

But I still had to offer to treat brunch to get the meeting. Robert always struggled for research investment and was in the midst of fundraising. Robert, by the way, is married to Marilyn vos Savant, reported by *Guinness Book of Records* as having the world's highest IQ. She is descended from famous Austrian physicist Ernst Mach, so no surprise there.

We met and ordered brunch, and we discussed the new Jarvik 2000 heart, which he showed me, and the challenge of an electronic control system. At another time we put the pump to use in my friends' kitchen at their apartment on John Street. I have a prototype of the Jarvik 2000 artificial heart in my possession to this day, but what I did not have at the time at the diner, when I searched for it, was my wallet. Sometimes al fresco is not all it is cracked up to be. Robert was gracious, if caught unawares, and it was one of those times when trying to explain that it was accidental made it sound worse. I've never been so embarrassed and I've never been game to suggest to Jon or Robert that I meet Marilyn. I'd need more EQ.

CHAPTER 8

DAD'S CHOOKS

My dad bred chooks. Lots of them. He was good at it. 'Chooks', the Australian jargon word for chickens, have been part of my life from the beginning. According to my grandmother Thelma, who was called Lil except by me, who called her Nan, my first sentence was, 'Chooky laid a googy.' A 'goog' being the equally absurd Australian vernacular for an egg.

My first three homes, in quick succession, were at Waddamana in the Tasmanian central highlands, and I have my first memories from the third of those, so I must have been about two years old, when the memory parts of the brain are developed enough to store them. As J.M. Barrie said, 'You always know after you are two. Two is the beginning of the end.' In my mind I can see the house from outside but not inside, with a rustic, slanting, unpainted, wooden car garage at the side, already old, and a paved street at the front that ended in a dirt turnaround area, as ours was the last house at the end of the street.

Waddamana was a beautiful, idyllic, alpine village; remote, self-contained, offering a dream childhood of freedom, adventure, exploration and simplicity for a child, even one as unrobust as I was. It was the location of the state government's first hydroelectric power station, at a time when politicians were statesmen and infrastructure projects for the long haul were the norm, unlike the career politicians of today whose only purpose, and talent, is to get elected and re-elected.

What I remember more was arriving at our next place, a small, abandoned army barrack, a hut really, at South Arm at the other end of the state. It was my fourth home and I would still have been only two or three years old (I talk more about South Arm later in another chapter). It was late at night, a day-long trek south across half of the state. I remember it was a lorry, and when we got there and stopped, all the hens in their cages on the back were dead. Whether it was exhaust fumes, anxiety or cold, I cannot possibly have thought about, but I remember Dad's devastation. Regardless, it means that we were already breeding chickens in Waddamana despite my having no recollection of them, and it was the subsequent memories at South Arm that I have of hundreds of chooks: hundreds of chickens, hundreds of eggs, and a single large round incubator with suspended heat lamp to hatch the fertilised eggs and seeing ingenious little yellow chickens unfailingly peck their way out. I have an old black and white photo of the hut from that time, and have visited there since, all modernised and expanded now, and described this history with the current owners.

At every house we lived in after that, the first thing Dad did was to construct a chook shed, a chicken coup, a hen run. As I grew up I participated, a carpenter's helper, holding joists and studs and noggins to block them as Dad built frames, me learning to construct modular walls flat on the ground first, pre-Ikea, then to upright them to see the hidden, physical 3D shapes emerge, like Michelangelo seeing a statue waiting to escape from a block of marble. Roosts up high, and laying boxes down low with their access from the outside through hinged hatches. The cereal smell of hot, humid laying mash we mixed as a regular chore and fed to the chooks each morning, sliding our hands under the warm hens and carefully stealing their eggs, talking to them.

My mum could make a guttural chicken noise from the back of her throat, a clucking, and she taught me to do it, the only other one in the family who could do it, and a favourite party trick. That and suspending a spoon off the end of my nose.

Even at the last house Dad lived in before he died he had chooks and so many eggs that he used to preserve them in agar agar, a sticky plant gelatine to seal the shell pores and prevent air and contaminants getting through, preventing the eggs from rotting. The stink of sulphur was ubiquitous. Funny that agar can prevent bacteria like that yet today it is the basis of most petri dish bases to actually grow bacteria, an idea from Fanny Hesse – another person named Fanny like my great grandmother. Unwanted roosters were doomed for the table. I remember back then that lamb was the cheapest meat, steak in the middle and chicken was a delicacy, exactly the opposite of today. We never ate fish. One of my most loved school teachers was Charles Wesley at Cosgrove High School. So much so that I talk about him more later on. He was my home class teacher and also taught us science. At age 13 I remember being fascinated by his lesson on the structure of an egg. We all knew an egg had a white part and a yellow part, but that was it as far as we were concerned. He pointed out the obvious: that the ovoid shape was so that the egg did not roll out of the nest but rolled in a circle and stayed inside the nest. And although we knew a chicken came from the egg, none of us related that to the fact that there was a little red spot which was the germ cell, to become the embryo if the egg is fertilised. We were too young to discuss fertilisation. And who would have thought that there was actually a spring inside to ensure the embryo always stayed towards the top of the egg as it rolled around, so that it collected the warmth of the mother's body? I wonder today if Dad knew all this about an egg's innards as he held one up to his faint Candler light to 'X-ray' what was going on inside.

When I was old enough to comprehend, I remember Dad saying his plan was always to make a business from this activity which was clearly more than a hobby. He loved it. But as good as he was at creating hundreds of chickens, his ability to convert them into a commercial success was absent. The South Arm chicken facility had two levels to cater to the exploding numbers; no more mass chicken deaths, I guess. It is a little like the dedicated people who love gardening for the sake of it rather than to prepare a low maintenance place, a retreat to eventually hang out and relax, read a book and sip a drink. My friend Sue lives in and for her garden, shady sun hat, fork and trowel in her gloved hands and ever-present knee mat. She never wants the gardening work to be finished. She loves the work. I have no memory of my dad ever seriously attempting any business enterprise. He was a worker. My friend John Reid once said there are two types of people: those who move pieces of the planet from one place to another, and those who tell people to move pieces of the planet from once place to another. Dad was a mover.

CHAPTER 9

BUILDING BOATS

Dad was a mover. But he was also a shaker. What became evident to me was that dad lived in a world he did not understand – a consequence of the working-class and unsophisticated family he was born into. Long after he is gone, I continue to analyse him and summarise him and study him in an attempt to understand him. So he resorted to a lot of shaking the system. Often with his fists. A very complex man. Maybe all dads are complex men. Everyone tells me I am just like my dad, so perhaps I am trying to understand myself.

While he understood chooks but not business, he understood carpentry in both practical and intellectual senses. His apprenticeship was in carpentry and he was a skilled builder. That is an insight to this man. I have vivid memories of Dad around the kitchen table: pen and paper and sketches of houses, with all the names of the parts, the cross-sections of sash windows, different roof designs, his tickle at my reaction to what for a kid were titillating names: a bastard file, that one a cross between a file and a rasp; or cranky grain, of wood that is impossible to plane smooth. I recall my delight in learning from Dad with his vast theoretical knowledge. His greater delight in teaching me.

A neighbour I overheard once said that she looked out her window one day and saw Dad starting a carport at one of our houses, and in a few days it was done, finished, painted, still there to this day. Sixty years later it's like it was built yesterday. The neighbour, Mrs Lobban, was impressed and I felt for the first time the pride of knowing from someone else, from independent opinion, that my dad was really good at his profession.

He put it to use. Every place we lived in was bought to renovate, and we lived in a permanent state of internal construction. An internal construction engine. It was only years later with the discovery of the effect of *Dermatophagoides*, the house dust mite, that had a debilitating effect on allergies, that the cause of my chronic and at times life-threatening asthma was the accumulated, infested dust from pulling up floors, demolishing walls and rebuilding these old houses. But a great learning curve for a hands-on, technically minded kid. I learned about tools, and their names, and functions, and proper use and storage and maintenance. Smells, wood shavings, essential oil aromas, identifying King Billy from Huon pine, and the excitement of creation, were our home life.

Meticulousness. Pride. Never taking a shortcut. One time he had to prepare and position a single piece of wall panelling with a hole cut out for a cupboard inserted between wall studs. He measured the dimensions for the hole, got sidetracked, and came back to the task asking me if the numbers he recited were correct. I might have said anything, and he cut the hole out of the middle of the large wall-sized panel. It was wrong, slightly oversized, and instead of being angry or frustrated he laughed.

He said to me, 'Never guess; always go back and double check. Let this be a lesson.'

Was he talking to me or himself?

For me it has been a lesson ever since. I was more upset than he was.

I said, 'What do we do now? The whole piece is wasted!'

He simply replied, 'It's easy. We put an architrave around the cupboard frame and hide the gap from the wrong cut.'

He was confident that he could recover from an inadvertent error. If it had been a business decision for a CEO, it would have been the equivalent of a contingency plan for recovery from a problem arising from a risky project. In general, I have a fall-back plan to most critical decisions I make, which requires deliberately assessing in advance the types of things that might go wrong, and the genesis of that approach comes from working with my dad. He could have been good at business.

Some of my early adult friends were into fishing and water and boats. I decided to build a boat, and using Dad's workshop under the house after hours I constructed a dinghy. I discovered that carpentry tools were not the right ones for boat building and neither Dad nor I had any idea about specialist tools like spokeshaves or jigsaws. I had bought plans and templates from somewhere, so had the design, but everything my dad had done in his life of woodwork was to ensure everything was absolutely straight, plumb, square, level, even, mitred, bevelled. Nothing in a boat is straight. It was fun using electric jigsaws and cutting out the curvaceous hull ribs to build those sensuous shapes, every rib different, unlike joinery or framing where identical multiples of each part are made, starting with the template which is always identified with a 'T' for 'template' drawn on it in pencil.

I did this work alone. Dad would come down later in the evening, though, intrigued, leaning against the door jamb, smoke in hand, legs and arms akilter, just watching. He just observed what I was doing, seeing the progress, not offering advice or suggestions, positive, supportive in his silence, just his presence. I think he was watching his son surpass his own woodworking skills but in a totally different direction. I have done house restoration and I will never be half the tradesperson my dad was. In that field he was a professional, an expert, a genius, a mover and a shaker. But he never built a boat.

Later, my cousin-in-law Leigh and I embarked on a project to build a real craft, a 21-foot New Zealand design called a Flareline by respected designer Richard Hartley. The skills from experimenting with the dinghy came to the fore as we constructed this beautiful, sleek, inboard-outboard power boat. Covered in marine ply, it was a dry boat because of the wave-diverting bow flares.

Never got to use them, though, both boats meeting ominous ends. The dinghy was lost in the Lonnavale house fire, and the other appropriated by Leigh. But I did buy another old inboard motor craft and enjoyed sailing it around the deep Derwent River estuary until it too had an ominous end, as it was crashed into a jetty at the seaside resort town of Dennes Point on Bruny Island by a friend, retiring it with its smashed bow. Not very dry after that.

Sister Dianna spent years with her partner Chuck, a carpenter, another man's man, so he and Dad gone on well. He was the leader of a local bikie gang, so he and Dad also had the odd problem because Dad was not comfortable with his daughter being in such a relationship. Chuck, though, was a talented boat builder whose passion was designing, building and competing in tunnel-hulled racing craft. The bikie gang was not a criminal one, and there were fabulous weekends of racing bikes around our properties at Lonnavale, partying and BBQs, reeking of testosterone and machoism, or weekends at racing carnivals with the smells of spray and racing fuel. Chuck and Di's good friends were Sean and Sandra Kelly, the world's best rock and roll dancers in my view. I attended Sean's funeral years later, and was surprised to have his son come up to tell me I was a hero in his family, apparently because of a prosthetic hand I had designed. The son said he built one from my design. I spent the whole afternoon with Sandra, whose funeral I was unable to attend a few years later because I was overseas. All of Australia's bike gangs travelled to Tasmania for Sean's funeral service; there were hundreds of bikes, requiring police escorts and traffic control. Amazingly, the chairman of a company I led in Queensland later was named Sean Kelly, and even more amazingly he was originally from Tasmania, and he said he knew of the other Sean.

All the real seafaring I did was on other people's boats: Zack's catamaran off Deer Isle in Maine, and Uwe Meffert's giant leisure craft in Hong Kong's South China Sea. Uwe is one of the most curious of all the people I worked with. Another European, tall, fair, hair cut in an artistic style; a totally debonaire Swiss-German. He invented and released the almost impossible to conceive of four and five-sided Rubik's cubes and dozens of other puzzles. It was the ingenuity of those cubes that tickled my fancy and I still have a sample. He was married, living in Hong Kong, but had been educated in Geelong, Australia, as well as other places. Truly worldly. I was surprised to receive a call from him one day – maybe because of our both being Australian – inviting me to come to Hong Kong. He said he had an opportunity to develop a hobby robot for the international toy market but needed someone like me to do it. It was to become my first international robotics venture – more about that later.

While in sweaty, dusty plastics factories in HK's Aberdeen Harbour, between lunches on the now defunct Jumbo floating restaurant, where the fish were hauled fresh and squirming from a well in the boat, I designed and built more robots, including presenting a Japanese-speaking robot to Mitsubishi in Japan, and the large hobby robot Elami for Japanese toy company Tomy Corporation. This project led eventually to Tomy's Omnibot, Mr Walker for Radio Shack and Chester for Commodore in America. Between the intense work, we went swimming in remote unpopulated parts of Hong Kong Island and had catered sailing picnics on his luxury yacht, diving off the gunwales into the ocean to swim to small, isolated beaches around the South China Sea. He'd ring his captain, who lived on board and maintained the boat, and by the time we reached the marina it was primed ready to go, fully stocked with food and wine for lunch.

All of the boats I'd built were motorised, and the only time I ever sailed myself with no motor was once in Maine, a little single mast yacht with a pull-up centre board. Zack kept it moored off his beach at Deer Isle and used it for fun around the bay. I would visit the house with Amy and we had permission to use the boat, but I was not confident, having never sailed by myself. Basically I knew the theory of sailing and had been on many boats when others were doing the work, but had never participated. I was a bit fearful of making a fool of myself, but I wanted to. Sail, I mean, not make a fool of myself. Eventually it got the better of me. With Amy and a picnic hamper for an on-board lunch, we swam out to the mooring, uncovered the tarps from the yacht and set everything ready to sail out of the narrows to the open bay.

There was a mild wind, enough to propel the boat without

causing consternation, and one of those high latitude sunny days which had ignited the idea of lunch at sea in the first place. Sailing out in the windward direction was a breeze, as they say, and I gained confidence. We anchored and ate. I knew that after lunch, if the wind direction did not change, which it did not, then I would have to tack back. That was the theoretical and more difficult part of sailing that had prevented my trying all those years. The concern was well placed. We spent about two hours trying to come back in, getting from the open water to the opening of the narrows, which is a neck of water between Deer Isle and the smaller Campbell's Island off shore in the middle of the bay. But at that point nothing worked and the boat simply moved back and forth parallel to the wind. I could not make it go into the narrows. Amy displayed amazing patience as I sweated at trying everything I could think of.

I was embarrassed and worried that our goings to-and-fro would look suspicious to someone observing, who might come to our rescue. That might have been necessary eventually, but I was not ready to give up yet. Perseverance. Persistence. Stupidity. I can do that! Finally, when we had stopped to rest and consider, both of us saw at the same time the drop keel sitting up out of its slot. I cannot remember now if I had forgotten to put it down or forgot to lock it in place, but for the whole trip we had been sailing without a centreboard, rudderless. Going out and floating around at sea at lunch time with no direction in mind was not enough to make it obvious to us. Finally, with the keel in place, the first try took us straight back into the mooring buoy. I made it look to any observers that this is what I had intended all day, but in truth I had had enough tacking experience to last me a lifetime.

Shortly after Zack Rosenfield graciously let me take the helm when he returned a leased catamaran to Rockport, never knowing what a poor sailor he had on board. His wife, Marydel, waited at the docks at Rockport to drive us back to Deer Isle, and the car was much more to my liking.

CHAPTER 10

WANKEL

When I was in California in 1987, I bought a Mazda RX7; a sleek, red, rotary engine sports car. There were family members who were into cars and bikes and racing them as I grew up. My Uncle Lionel Hart rode motorbikes in the Globe of Death at fairgrounds. Easy-going, confident, fearless, fair-haired, nicknamed 'Snowy'; amazing to have someone like this in my family and it means more to me now as an adult than it ever did back then. Back then he was just my affable Uncle Lionel, distinctive because of his permanent limp and a missing finger. Short but powerfully built Uncle Cedric Calvert, nicknamed 'Bull', was one of those magical mechanics. He could make and fix anything, his hard-working hands belying his landed gentry background. Others rode open wheeler racing cars. My Uncle David Branch, who was Tasmania's state go-cart champion and second place in Victoria, built all of his own vehicles. My earliest best friend Eddie Sauer, a science teacher and high school laboratory manager, and I, in a garden shed in his backyard on the aptly named Waterworks Road, converted an old Flat-4 VW engine to run on 80 per cent water and 20 per cent petrol, and then modified his FJ Holden work car to do the same on a daily basis on the open road. So, playing around with and having an exposure to, an interest in, and a knowledge of cars and engines and mechanics was always part of my upbringing. Today when I look under the hood of a car I recognise nothing, but there was a time when I could diagnose any problem and rebuild anything on an engine, or almost any mechanical device for that matter.

I was fascinated by the development of the rotary engine invented in Germany by Felix Wankel, unbelievably much earlier, in 1924. I love innovation particularly when it is aligned with elegance. The Wankel engine was innovative, though it took a long time for the engine to become elegant. In the mid-1960s I followed the story of the NSU Prinz Spider, fitted with the prototype of the Wankel engine. I bored my Uncle Lionel with my conversations about it. I could not understand why it was not so interesting for him. At that age I understood the principle, the design, the mechanism. I did not understand as much then about materials, refractory metals, friction, heat, seals, thermodynamic efficiency, and all the problems that beset the engine, but who cares? A good idea is a good idea and should be persevered with, something that was to become a sort of modus operandi for me.

Because the engines rotated so fast, in fact needed a regulator to limit the speed, and because the outer part rubs against the outer metal block, a seal is required, and the technology for these with heat and friction challenges was one of the limiting factors. Eventually the perseverance paid off, with Mazda manufacturing successful Wankel engines for its RX7. And now I had one.

I didn't know then that Wankel was a Nazi sympathiser. At that age, I knew nothing of Nazis. After World War II, Felix Wankel was jailed, released, and banned from more inventing, but like many Nazis post-war, was resilient. He became successful and even licensed his first design to Curtiss-Wright in the USA. NSU became Audi.

Thirty years later a key meeting had been set up for me with electrical tool and appliance manufacturer Black+Decker. I was promoting my robot vacuum cleaner design. My little red engine was in wintery New York, driven across country from California, where I had bought it from friends. I had parked it, as one does in NYC, bumper to bumper, near West 86th Street where I was staying with friend Faye. Parking space is at a premium and it is usual for cars to nudge into spaces with as little as a foot on each side, deliberately edging until they actually touch the cars in front and behind. Parking by feel is what they call it and it is accepted. After a lot of searching I had found a spot on a side street outside a building construction site. When I walked to start my car on this snowed-in morning, I saw that overnight my car had been slammed into, presumably by a construction vehicle sliding in the ice outside the work site. Workers on the site adjacent were staring down at me wondering how I would react. The driver's side door was caved in, but otherwise the vehicle was drivable, and I was late for my appointment. So I did what every foreigner does in a strange country, nothing, and drove out to Towsen in Maryland for my meeting with Black+Decker. I had to climb in and out by the passenger side.

By then I had done similar product presentations with vacuum cleaner companies around the world, so was not worried ahead of the meeting, which went well enough, although nothing came of the presentation.

Out of the blue, though, shortly afterwards, I received a call from someone with the improbable name of Tory Orzeck. Tory was calling from General Electric Plastics, GEP, in Pittsfield, Massachusetts, and he wanted me to submit a written proposal for developing a robot vacuum cleaner for them. What are the chances? I researched GE Plastics and Pittsfield, a small outpost of a town in rural western Massachusetts, almost into Albany in New York state, and the location also of GE's transformer division. Black+Decker was important enough, but GE was something else. This time I really was a bit apprehensive, so I spent more time thinking about what to put into my proposal. I decided to go all out. My submission was all-encompassing. I wrote up the whole technology specification and project outline, costing, time, resources, and it worked; I got the contract. It was a coup. It is always a complex choice to decide how much technology to disclose to a potential client. I thought, if GE is contacting someone like me, then they cannot do it themselves, so it would be safe to detail more of the technology. It is possible to make the reader of a technical description feel like it would be easy to do themselves, until they try it and discover that there is a need for critical know-how that is not disclosed in the descriptions.

On my first day there after starting the project, chatting with two telephone receptionists at the front desk, I could tell that they were talking on a phone to a robotics company about this same project. When I asked, they said it was a company called Denning, which I knew well and will write more about later. Denning was begging for GEP to cancel my contract so that they could do the project themselves, for free. I asked why they would do that and was told for the prestige of having a GE affiliation.

My friend Faye came up by train from New York to visit me in Pittsfield. We dined at the Misty Moonlight Café, where every tenth song played on the jukebox was the Jerry Wallace song 'In the Misty Moonlight'. 'Misty Moonlight' was the first 45 RPM vinyl record I bought, and later Faye gave me a copy as a surprise gift, which is a prized possession in my collection in Tasmania.

At some point I discussed the Denning call with Tory, wanting to know more about why GE chose me, a much smaller company than the very visible NASDAQ-listed Denning. He said it was because I disclosed enough for them to determine my technology was real and worked, something no other tenderer was prepared to do, so they were able to convince the decision-makers at GEP.

This was a very expensive project and GEP injected huge resources into the project, not just engaging me. Their whole design team was committed to the project.

So as a team we built Florbot, a prototype released at the subsequent Electronics Show in Colorado that year. Here I was holed up in a motel in this small town of Pittsfield in western Massachusetts, working in a side room at the GEP offices, while Tory, a brilliant project leader, designed the body and chassis, plastics engineers making the prototype body, to take the robotics components I added into it. The first cyclone vacuum cleaning systems were appearing, destined to replace the older filter bag systems that had been around since day dot, and we used one in the cleaning and filtering sections, a revolution almost as significant as the robotic navigation. Sometime later the robotic technology from this and other projects was presented and disclosed to the three executives of iRobot in Boston, which I believe led more or less directly to the Roomba. So that is Florbot's legacy.

There was another legacy. In fact, it was a shock, and illustrative of the American corporate networks at the time, to discover that hidden in a written testimonial from Robert Williams, the Manager for Consumer Marketing Programs at GE Plastics, with invaluable compliments on my work for GEP on Florbot, was a sentence at the very end saying that his friend and colleague at Black+Decker had also promoted me to the GEP Florbot project. There was never a hint to me of that connection, which is how GEP became aware of me. In a way the Black+Decker presentation turned out to be a success story after all and it taught me to always be careful what you say: never burn bridges.

GEP took its promotion of products seriously. The filming was done in their display house. At nighttime, but with flood lights outside the windows to make it look like daylight inside. The house had the first examples I had seen of curtainless LCD darkening technology. Press a switch and the windows became opaque. A special hand model was hired – first time I knew of such things – and her perfect hands are in all the images except one in the brochure, which are mine. In the lobby of the GEP offices was a demonstration car. At the front and rear were prototype plastic bumper bars, replacing the usual metal ones. I remember saying to Tory that no one would put a plastic bumper on a car. So much for that.

Just before the COVID-19 pandemic I visited the incomparable German Technology Museum in science-heavy Munich, where I stay a lot with my German friend Sabine Zenker. Of interest were the V1 and V2 World War II rockets, which I relished reading about because I'm a baby boomer and had read so much post-war material during my youth, leading to more instances of Nazism being overlooked when Russia and America and other countries wanted to recruit rocket know-how after the war. But by surprise, I turned a corner in the meandering halls of the museum to see sitting in front of me the handmade, first workshop prototypes of Mr Wankel's rotary engine. Talk about come full cycle, pun intended.

Another clever mechanical designer and entrepreneur, in my view, is Ralph Sarich, an Australian who solved many of the inherent design problems of a super-fast spinning rotary engine, with his so-called orbital engine. Instead of rotating like a wheel, it oscillated with a motion a bit like someone wiping a window with a rag. So, it did not actually spin, it moved around in an eccentric orbit. The irregular shape of the block created the compression as volume decreased in the chambers necessary for combustion. So, there was not the heat and friction and oil seal problems of the Wankel design. Sarich, I think, was a bit like my father: impatient, intolerant of fools, full of good ideas and could create the raw product, but not so much the business, which is all about relationships.

But my General Electric story is not finished. When Jeff Immelt took over as CEO of GE, I called him up. Well, not quite as simply as that. My friend Sabine in Solingen worked for GE Medical Systems in Europe and I called her up.

I asked, 'Do you know how I could contact Jeff Immelt?'

She said, 'Yes, he is standing beside me at this very moment.'

'Great,' I said, 'Can I speak with him?'

She laughed and said, 'Of course not. Do you want me to lose my job?' But she said she knew his personal secretary and if I contacted her, she would ensure my email would get through to him. Personal secretaries being the gatekeepers they are. I did, and received an invitation to present to the executives at GE head office, which was still in Fairfield, Connecticut, back then.

My hook to get an audience with Jeff Immelt was a line in my email. In marketing, a 'hook' is a comment designed to make it difficult for the other party to say no to your request. It is often used by politicians deliberately incorrectly, linking something they want but which is not popular or difficult to achieve with something impossible to say no to. 'Of course we must build these multistorey apartments; do you want babies to grow up malnourished?' That is a sophism. Mine was a use of identity logic, a comparative reference to one's reputation. Robotics is always topical but an untapped opportunity, and I said that Jack Welch's legacy was plastics and that Jeff's could be robotics. That GE could become the world's leading robotics company. Hard to ignore a tease like that. Are you going to say, no, you don't want to be the biggest robotics company? I was also very aware that the acquisition strategy at GE was to only pursue companies first or second in their industry, so to be number one had a certain ring to it. That I had already worked with GE on Florbot was a plus.

At GE I met with Scott Donnelly, VP of Global R&D. My best friend, Zack Rosenfield, drove me up from New York. For all his worldliness and experience in a top architectural firm in New York, Zack was excited about tagging along to visit GE with me, acting as part of my team.

From the start, things did not go well. Despite being a technology company none of GE's audio-visual equipment worked, and instead of projecting my PowerPoint to the overhead screen I had to run it on a small laptop screen, so to my mind the oomph was lost. But the meeting was a success. Afterwards we had a guest lunch: sandwiches on a trestle table in the corridor outside the meeting room.

Another time, when I was Invited Visiting Scientist at Carnegie Mellon University in Pittsburgh, I did some robotics projects for General Motors, Hilton Hotels, JPL and NASA. In fact, the NASA project was exciting enough that they created a page on their website suggesting that my robot had been developed by them, an autonomous industrial cleanroom cleaning robot modified from our product called RoboScrub, which was manufactured and sold by Windsor in Denver. The robot had to clean the Shuttle loading bay without any contamination by machine or man that might affect the delicate payload satellite electronics. No dust, no splashes, no oil droplets, no scurf. It was interesting to tackle NASA and have the web page removed.

Being Australian, a meeting at the Space Shuttle launch site in Florida took some significant security clearance, but following the presentation I was also invited to lunch. We went to the staff canteen, lined up and waited. I paid for my meal myself, and we sat on basic metal tables eating a quick worker's lunch with the top executives of this division of the world's largest space agency.

Europe is not America. By contrast, when I had meetings with Miele, in Bielefeld, Germany, a large high quality household appliance manufacturer, after each meeting we went to the exclusive executive dining room, with the whole executive team, and had white gloved catered banquets. The 27 member European Union as a bloc is larger than the USA, 450 million people and an economy that is on a par with the USA's. But Europe has an ancient tradition of elegance and distinction and sophistication that is not evident in American companies.

They say that imitation is the sincerest form of flattery, but not in the commercial world. Another time a Canadian company released an article in an international plastics trade journal, claiming to have developed Florbot for GE. I was in New York and forget how I came across the magazine. Livid and confused, I discovered the journalist and magazine offices were up the road in the Times Building. I walked up and confronted the journalist, and within the pages of the next month's issue had a retraction and a feature about the real development of Florbot. Only in New York, centre of everything, could that have happened so readily.

One day I was driving my RX7 across country to an industrial project when it broke down in Kentucky outside Paducah. That is funny enough because of a Tom T Hall song about how it never rained in Paducah. It is also funny how things happen to me in remote rural towns. I managed to limp my smoking car to a garage off the interstate turnpike and left it there to be repaired. Paducah is on the Ohio River which comes from Pittsburgh and is famous for making bricks and enriched uranium. This part of America's heartland had few Japanese cars, no rotary engines and no mechanics to repair them. My Australian accent did not help either. I needed to rent a car to continue my road trip to Tennessee, to meet with Owens Corning at Jackson and hoping to bump into rock and roll pioneering icon Carl Perkins at his restaurant Blue Suede Shoes across from my motel. A couple of my workers had already met him and acquired autographed CDs. In Paducah, a town of 25,000, I was lucky that there is an airport, the only place to rent a car. The garage owner kindly offered for a junior apprentice to drive me out there. On the way, he asked me where I was from. Tongue-in-cheek I said Pittsburgh. He looked at me with a satisfied smile and said that he knew they spoke differently up there. The apprentice had probably never been out of Paducah. It was years before I eventually had a chance to go back to Paducah, after completing projects outside of America. When I did, the old car, partly stripped for parts, dent still in the side door, was still there in the back of the yard covered in dust. It was never going to be fixed.

Mim and I produced a radio program of country music when I was turning around a radio station in Tasmania, one of the perks of being the boss, and she played the Tom T Hall song while relating the RX7 story to the listeners. She had driven the RX7 many times when we were in Sunnyvale California. Tom complained fatalistically that it rained in every town except Paducah and then it rained in Paducah too.

In late 1993 I took over management of Denning, the company that had called GEP and tried to take the project away from me, and turned it around from insolvency and impending liquidation. Today I look back on Denning as my last robotics project and my first company turnaround, a change in direction in my career and life.

CHAPTER 11

CHINON

It is a wonder I have not lost friends because I take so long at museums. Maybe I have. Body language tells me friends are sometimes impatient, frustrated, trying hard to be polite and accommodate me, but so far I think they have stuck with me. I become totally engrossed with art and architecture, history and natural history, science and technology, so a good museum or exhibit, art collection, display, castle, gallery, historical site, natural wonder, captures me.

The area called Alsace, currently part of France, is one of those border regions that sometimes was claimed by Germany, sometimes by France, so has a mixed cultural inclination, including its language, Alsatian. When the word 'German' was out of fashion during the first world war, German shepherd dogs were renamed Alsatians. Alsace is in a lush region that is known as Lorraine, a word that is Germanic but sounds and looks French. My sister Dianna and friend Jon Jarvik drove with me from Caen to Antibes, from the very north to the very south of France, and stayed overnight in Alsace before continuing on the gravity-defying, cantilevered, almost floating freeways bordering Switzerland, terrifying each other with our mountain-road, snake-winding, hairpin bend driving skills. Relieved to reach Lucerne intact, for another stay where Jon had arranged to meet research colleagues, charmed by the rocketing water fountain in the eponymous lake, then on to the unbelievably blue Mediterranean. Somewhat earlier in Alsace, in February 1429, a very young peasant girl from the town of Domrémy in this part of France did something that resulted in the townspeople not having to pay any taxes. She had a vision from God in which she was instructed to raise an army to fight not the Germans, for a change, but the English who were invading France at that time. Her name was Joan.

It was a time of disenfranchisement for the French, with the English King Henry V in command of most of France, a consequence of the Hundred Years' War, and the new-to-the-throne King Charles VII of France was really king only of Bourges, a backwater along the Loire Valley in central France. He had a castle in the nearby town of Chinon, famous for the emerging pre-science of alchemy, but which when I visited was awash with cool, hillside caves filled with maturing local cheeses, also a sort of alchemy.

Charles VII wanted to be coronated properly in Reims as the true King of France, and Joan was told by God to help him. God was always telling you to help others back then. Paris was out of the question for a coronation because it was in Henry's control and besides, Reims was the earlier traditional place for such ceremonies. Joan was to raise an army to fight the English. The trouble was that only the nobility had armies in those days so she had to get permission, which meant a road trip to Chinon. Charles had heard of this peasant girl, was sceptical of her claims of Godly visions, but to hedge his bets he decided to test her in case it was true. He agreed to receive her at his court. People were always hedging their bets against God in those days.

The story, as I learned it at school, was that for the reception the King hid among his courtiers and placed an imposter on the throne. Somewhat like the insane megalomaniacs in control of authoritarian countries today who often put lookalikes in harm's way to avoid assassination. Interesting how many of them have funny moustaches: Hitler, Stalin, Hussein. Skinny, severe-looking King Charles VII of France did not have a moustache and was not insane, just not a betting man. In those days, unless you were a member of the court, it would be unlikely that you knew what the king looked like anyway, but Joan walked up to the imposter on the throne, and instantly said that he was not the king. She turned around and espied the real king hiding in the crowd, then came over and bowed to him. That was convincing enough, and Charles allowed Joan to raise an army.

There was more testing, though, as kings are prone to do, and he sent Joan, a 17-year-old naïf, up the road to Orléans first, where the French were struggling against the English. Within nine days Joan had remotivated the troops and the English abandoned the Siege of Orléans. Joan pursued the English, defeating them across France, with the result of King Charles VII of Bourges becoming King Charles VII of France at a veritable coronation in Reims. It was as he always wanted, now with Joan again in the King's court, but this time alongside him in a place of honour. The king was so thrilled that he ordered that no residents of Joan's hometown would ever have to pay taxes again, and they never did; that is, until Napoleon took over and changed the rules.

On my road trip through the Loire Valley I stopped at the castle at Chinon, and realised I was looking at the very room where the first test of Joan by King Charles VII of France took place. At least the remnants of the room, with holes in the wall to take floor joists of the second-storey room where this supposedly happened, and a fireplace, where one can imagine the king warming himself ahead of Joan's entrance. It was here that I recall my friends who were travelling with me displaying otherworldly patience as they waited for me to check out every little nook and cranny and revisiting the story as we learned it. Mind you, this tale is 500 years old and could be half-history half-myth.

We all know the rest of the story. Joan and her army were subsequently defeated in Paris and in one of the cruellest acts ever recorded, the English burned her to death in Rouen, aged 19.

In our modern times we are inundated with sham trials, not just from sham sovereignties, but everywhere, like the Lindy Chamberlain one in Australia about the dingo that took her baby, or the dozens of prisoners on death row in the USA lucky enough to have their innocence proven before they are murdered by the state. We're almost to the point of being immune to such atrocities, but the trial of Joan is an example of the most callous, most sinister, most corrupt, most politically charged trial one could imagine.

Joan called herself Joan la Pucelle, which means Joan the Maiden, maiden a euphemism for virgin; but the king called her Joan d'Ay de Domrémy. Her father's name may have been Darc or Tarc, so eventually her name became Joan Darc, or as we know it, Joan d'Arc meaning Joan of Arc.

CHAPTER 12

FALAISE

People are often surprised when I say that English is a Germanic language, related to German, Swedish, Danish, even Dutch although they hate to admit their language might in any way be related to German. Unlike the other Germanic languages, which are reasonably pure, at least until recent times, English is highly adulterated.

Through the Académie Française, France does everything to prevent their language from diffusion by other languages, especially English, so French is in some ways more reminiscent of Latin, the so-called dead language, except in France there is a government department to create new words as needed, else it would become stagnant too. France refuses to accept foreign words and expressions, even words for creations by inventors from other countries, where the creator's term is usually adopted by everyone else. So, a computer, not invented by France, is called a computer everywhere, but in France they are ordinateurs. Vacuum cleaners are aspirateurs. Of course, on the street there is more flexibility and tolerance.

With the domination of the USA and its influence and power through the Internet and movies and music, and since the 1990s with the adoption of a common language by the European Union, English has become the global language, much to the chagrin of the French. Chagrin in French is the word dépit, despite chagrin being originally a French word. Go figure.

There is a reason behind all of this. In 1066 a Frenchman decided he wanted England and sailed across the English Channel from the Normandy Coast to get it, in the opposite direction from the D-Day landing of World War II, which was from England to Normandy. He landed at Hastings on the southern coast of England and after killing the king and defeating the army, set himself up to run the country. The first thing he did was establish his native tongue of French as the official language. So England spoke French for a few hundred years. But like the French street language that makes its own way, street English persisted, so that when the French influence finally waned, the old English resurfaced.

Then something amazing happened. It turned out that there were now two words for everything in England. For example, vegetable in French is legume, while in English it is, well, vegetable. This surfeit of words led to a richness not enjoyed by other languages. English was able to adopt the second word to add nuance or subtleness. So, in English, legumes became a specific category of vegetables: peas and beans. Work is work, but the French 'travail' becomes hard work or agony. House is house, while mansion from the French 'maison' is a big house. Woman is woman and a dame is, well, you know. All of this is because of that Frenchman, William the Conqueror, Duke of Normandy, who a thousand years ago decided to invade England.

Right in the middle of Normandy, in the middle of the city of Caen, where I lived and worked with Moulinex, close to the English Channel coastline from where he embarked, is a large thousand-year-old fort-like castle called Château de Caen. Built in 1060, the portal is a recognisable gateway, a drawbridge over the moat with its gatehouse over the entrance, and indeed there is a moat around the castle to this day, albeit an empty one nowadays. The massive ramparts cover over 5 acres of land, making it one of Europe's largest castles, and although the keep and residence of William are gone, the archaeological foundations are visible. It is a protected historical site. It is the home of William the Conqueror, and we enjoyed irreverently calling it Bill's Pad while we were there. For an Australian, with our European settlement history only going back a little more than 200 years, to see and visit and walk around a place that we learn of only in history lessons at school, but could never contemplate in reality, was mind blowing. I remember after one of my visits with friends to the interior of the castle, lingering and walking around the outside, and looking at and thinking about the first stones to be laid around the perimeter defence wall, and imagining stonemasons and skilled workers cutting and laying the foundation blocks, mixing mortar, measuring and surveying, just as they would today. I looked at a mortar joint on the first stonework layer and said to myself that someone mixed sand and lime and water in a bucket one day and laid those stones.

And it is not just the castle in Caen; the city holds prime examples of medieval churches including the Abbaye-aux-Hommes and Abbaye aux Dames. For me they were awe-inspiring buildings to wander through, considering the architecture, the religious relics in the nooks and crannies down the side of the Nave. Abbey-aux-Hommes, or its proper name Abbey of Saint-Étienne, where a marker sits on William's grave, is Romanesque, a dark, brooding, bulky structure unlike the later and better-known Notre Dame in Paris, with its thinner walls and large colourful glazed windows letting in light. Architectural technology at the time of William required massive walls with the strength to bolster up the weight of the structure, with few window holes because they only add weakness. They were, therefore, simple practical designs showing little to no style, with roof-top castellations, and narrow archer slots in the walls the only openings.

It would be a couple of hundred years before the technique of building lighter walls and adding buttresses to shore up the sides evolved, interestingly from this same region of France. A buttress is essentially the same as propping a brace to the side of a wall to prevent it collapsing. The method allowed large open windows for the first time, seen in the beautiful leaded glassworks of later churches, and the larger buttresses lent themselves to design and style as the move to the artistic Renaissance generated an interest in grace and elegance. They became known as flying buttresses. Flying, like multiple wings alongside the buildings, having a life of their own. Alcoves, niches, grottoes and other practical use could be made of the interstitial spaces. It is strange that today in English the word for these newer structures, Gothic, reeks of blackness and darkness, yet in reality it allowed the churches to be open and airy and full of light for the first time.

William the Conqueror, before he was a conqueror, had the ignominious name of William the Bastard. Maybe the desire to remove that name was part of his motivation to take on something more? I doubt it, because he already had a pretty strong hold on power by then. Bastard or not he had a father who lived in a nearby town. In the beginning, Duke Robert the First, a bachelor, was looking out of the window of his castle one day, watching the washerwomen doing their laundry in the creek that runs by his castle, and saw a beautiful woman who he invited up to his room, perhaps to see his etchings. At least that was what I learned at school. Alternatively, it is also recorded that he was looking down from the roof of the castle at the women doing their leather dyeing. It doesn't matter, because Herleva, which was her name, stopped whatever she was doing to meet the Duke and inevitably she gave birth to William, still out of wedlock. That town where this story takes place is called Falaise, a stone's throw from Caen, and the castle is Château de Falaise, although the one I have roamed around, visualising what Duke Robert was looking at, is a slightly later one built on the same site, much smaller than William's castle in Caen. The creek, if that is the true story, is still there. Falaise means cliff, so in English we'd probably call a similar town Clifton. Most castles were built on some geographical outcrop like a tor or a cliff for additional protection, as were Bill's and Bob's. I wandered for hours in the footsteps of William the Conqueror and his mum and dad.

On the other side of Caen is another town called Bayeux, and in a dedicated, darkened, temperature and humidity-controlled museum, is a 230-foot-long handwoven tapestry, embroidered in England, and depicting the Norman Conquest of England by William, like a serial graphic novel. I was fascinated by the graffiti and enhancements to the tapestry done by unknown vandals over the centuries, some of it pornographic. Vandals is a Germanic word from the name of the very first of the German tribes.

It is interesting, to me at least, that William the Conqueror was descended from Vikings, those invaders from the north who just a couple hundred years earlier attacked and occupied the north of France, hence the name Normandy, meaning northern people, and who spoke older Danish and Scandinavian languages, all Germanic, before integrating and becoming farmers speaking French.

Another time I visited the north-western part of France, Brittany, or Breton, which means 'little Britain', because of the early English settlers who travelled down to there from Devon, or Cornwall or South Wales, a sort of New South Wales, taking over the Armorican Peninsula. This was after the fall of Rome, and in time those early settlements merged to form the Dutchy of Brittany, and still today the French dialect there is a strange hybrid of old French and old English. I was there to see the Alignments at Carnac, straight rows of menhirs that are a linear version of the round ones like Stonehenge, structures that were thousands of years old at the time the pyramids were being built. Probably set up to point to the sun at the solstice, but no one knows for sure. From the ground under one of those megaliths I stole a small jagged rock, which may or may not have fallen off this prehistoric monument. Then I climbed the burial mounds and ventured into a barrow, seeing the lintel structure that created the menhir, and visited the multichambered cairns. All great words. We're lucky to have these structures, because at one time local farmers who owned the land they were on removed some of them or used the stone for other purposes.

What I think of as I write this is what an amazing adventure I'd had up till then: from hunting in the wilderness of Tasmania, to digging holes looking for ores in the Australian outback, to dating beautiful women in New York called Maggie or Faye or Virginia or Melanie, to sailing in the world's seven seas, to creating alongside some of the world's smartest scientists, to experiencing neolithic remnants in Europe.

CHAPTER 13

ROBOTS

Becoming a roboticist was inevitable for me.

Mechanics were all around me growing up. Uncle Cedric was a farmer from a farming family, but his passion was mechanisms and his farm shed was a massive workshop with every machine shop tool imaginable. In later life Uncle Cedric ran automobile garages and I used them to work on my first cars. I used his tools for other things. In my early electronics days, when still a teenager, I made radios, all with vacuum tubes, or valves as we called them, which required round holes to be cut through the metal chassis to mount the Bakelite valve sockets in the base, and it was Uncle Cedric's metal hole cutters that I used. I remember him keenly observing me making a radio, something outside of his experience, fascinated, as my dad was when I made the dinghy.

Uncle Lionel Hart was a rigger and a welder of heavy equipment and also a talented mechanic, but more remarkably he was a fearless daredevil like Evel Knievel. He rode motorbikes in the Globe of Death at fairgrounds.

Uncle David Branch was like Cedric, a homemade mechanic who could repair anything, fabricate anything, solve any mechanical challenge. They all lived in oil and grease. All of them competed in racing cars at times.

Dad was a carpenter, a different breed. He was a competent mechanic as well, but more at home with a saw than a spanner. Many of Dad's other friends, like petite Jimmy Brown, mostly bred and trained and raced horses, and some of them were different types of riggers, 'rigging' the races, giving us furtive tips like 'Number 7 in race 5'. They rarely bared fruit.

Sometimes the mechanical and woodworking skills of Uncle Cedric and Dad, who were the closest of friends, combined. In the mid-1950s Dad came home with a new car; in fact, the first car I remember us having, a Standard Vanguard, manufactured by the same company that made Ferguson tractors, powered by a bulky 4-cylinder tractor engine. It could not have been Dad's first vehicle, because we arrived at South Arm with the chooks in a lorry that I presume was his, but our first family car. South Arm was a fertile peninsula of land jutting into the massive Derwent River where Uncle Cedric, part of the landed gentry, had a large potato farm and Dad worked with him. Our Vanguard was a ute, short Aussie term for utility, suitable for a farm, the Australian version of a pick-up truck but more comfortable. It had a sedan front with an integrated utility tray with panelling along the sides to the rear and a tailgate. The closest in the USA is an El Camino. Dad needed a canopy to cover the tray, so they decide to make it themselves in Cedric's fully stocked workshop at South Arm. To mould the plywood to a curved U-shape, they first steamed the wood under hessian matting made from sacks until it was soft and pliable from steam drawn through hoses from a furnace boiling water, then rolled it to a metal frame to get the desired shape. I have always had a favourite memory of that project in progress at Uncle Cedric's spacious workshop: plywood flat along the floor and steaming as if capping a hot spring. Not old enough then to be aware of the inventiveness and ingenuity of that day. I remember also axles and pulleys and belts running along the rear of the workshop, rotating all the heavy-duty equipment like lathes, drills, saws and mills, compressors and boilers, from a single power source. Years later I saw the same type of old-style machine shop in Solingen in Germany where they made knives and scissors. A pre-war factory preserved as a museum. I wish my Uncle Cedric's workshop was still there, preserved as a museum. I wish my uncles and Dad were still here.

As the number of kids in our family expanded, we would sit protected from the elements in the back of the Vanguard under this canopy on blankets and pillows as we all rolled off to Waddamana for weekends and holidays of adventure.

Every Christmas and birthday, my gifts included Meccano sets; Erector sets in the USA. I loved them. I quickly outgrew constructing cranes and trains from instructions in the kit, and began making my own designs. I recall a vehicle, car or truck I do not remember, maybe a ute, and it having steering, which was not unusual, but I gave it a steering mechanism like a car's rack and pinion. A miniature steering wheel actually turned the front wheels. I almost definitely had help from one of the uncles, as the extended family usually ended up at Nan's place at Waddamana for Christmas, but I think of it as something I did. My mechanical interest progressed to gramophones, those wind-up, spring-driven mechanical devices with large sound horns and stainless-steel needles on a fast spinning 78 RPM shellac disc, preceding electronic record players. Scratchy sounds. Uncle Lionel and Dad removed the coil spring whenever I overwound it breaking the end, using the flame from the gas stovetop in the kitchen and tools to shape a new end so it could be used again. A miniature blacksmith's furnace and anvil. I watched and learned and participated.

As I grew up, I could repair an engine or mechanical device like a kitchen appliance, so my mechanical skills were therefore not bad, but nothing like my uncles'. My ability proved to be suited to more creative endeavours. At one stage I lived in Ponsonby, in the suburbs of Auckland, New Zealand, and was working with Bob Duncan at his electronics repair shop, Bear Electronics, in Ellerslie. I'd use the fantastically named Kyber Pass Road and

Lady's Mile to get to work. I'd bought a cheap, used Austin 1800, dirty brown colour, that burned more oil than petrol. When repair jobs were slow, we'd select something from the back room which held a collection of rejected repairs, those too expensive or too difficult to repair, and try again to fix them to resell for extra cash. It became a competition as we all diagnosed and repaired the radios, televisions and record players, trying to outdo each other. Eventually there was only an automatic record player left, the type where a dozen or so records are place on a spindle and played automatically one after the other. I asked Bob about its history. He said everyone had had a go at it, but it could not be fixed. I said, 'I can do that.' It turned out that the complex player mechanism had been pulled apart, and what I started with was a box of strangely shaped pieces, like Meccano, which I managed readily enough to reassemble. That was already impressive enough to my intrigued observers, but it was not functional even though everything was in place. The comments were that I had reassembled the parts incorrectly, but it was clear to me that there was a missing piece. We searched through the storeroom but there were no more pieces. No one had recognised that problem, which is why it would never play. Using my imagination, I worked out what the missing piece should look like, what function it would have had, then fabricated one from a bit of plate metal, a strange three-sided lever arrangement, and made the player work. This visualisation of problems and solutions, in a mechanical sense, is a key to how my brain works, I think.

A similar story is from many years later when at Jon Jarvik's in Pittsburgh I was boasting about these skills and he said he had just such a player in the attic that also did not work. Records always got stuck and never dropped. Did I want to fix that? In fact, he was calling my bluff. Of course: 'I can do that.' The central spindle in these machines had a slider in a slot in the shaft that looked a bit like a nail file. It had an edge sticking out proud that held the records up, and at the end of playing a record, the slider inserted itself flush into the slot in the shaft to let the next record fall down under gravity. The slider had become bent and the friction rubbing against the sides of the slot was enough to prevent it going all the way into the slot. The next record would never fall down, and the same record on the platter played over and over. It was the difference between knowing the basics of machines – wheels, levers, pulleys, screws – versus the real world of materials: friction, heat, gravity, expansion. Jon was, and is, impressed, especially after I demonstrated invisible mending on two of his jackets that had holes in them. I used to darn my own socks too, but now in this throwaway society you can buy a six-pack of tube socks off a cart on a dusty New York street for less than a spool of cotton.

The turning point for me was reading a series of articles called *Bionics* in a British hobby magazine. It may have been called *Practical Electronics*. I was maybe 17. The articles were instructions for making electronic robotic mice, similar to William Grey Walter's tortoises. One of the last issues in the *Bionics* series demonstrated self-recognition, through simple light and photo sensor circuits, a sort of primitive insect vision. Two of the robots would react if they saw each other. It fascinated me. I felt that I could make a better system, which gave me an interest in vision systems, which led later to my submission on vision in the students science talent quest. The submission, 'A Trichotomy on Vision', introduced pattern recognition using neural nets. I know now that it was the germination of my transition from an interest in physical sciences to biological sciences as I studied and learned about vision systems in all manner of organisms. It also led to my side work to test the idea with Dr Ian Newton at the University of Tasmania when I was a student at medical school and he was my Medical Biophysics lecturer. By then I had developed the early pattern recognition concepts from the Trichotomy on Vision to a full-blown AI system to recognise hand-drawn patterns. Before personal computers, after hours we commandeered and programmed the university's mainframe computer to run simulation programs, to assess the theoretical pattern recognition's ability to recognise handwritten characters. The computer normally took care of student files and courses.

In the history of automata pre-1948 and especially during the 1800s, there were examples of earlier automaton mechanisms – fascinating and ingenious clockwork things writing letters, drawing pictures or playing music usually. Impressive but limited. In my view, Grey Walter was easily the world's first roboticist. Grey Walter created robotic machines that were dubbed tortoises. They displayed phototaxis, an attraction to light. When they saw a light, they homed in on it, leading to a hutch to recharge their batteries, years ahead of the Johns Hopkins Beast. He is important to me for his robotics, but he was a pioneer in much more, including, amazingly, his early work on imaging tomography, the technology behind CAT scans, which have revolutionised medicine.

Then, while still at university, something tragic occurred. A close friend of mine who was also a student in Dr Newman's classes went home to Malaysia for the Christmas break and returned with a leg amputated because of a motorbike accident. He approached me and asked if I could devise an artificial leg for him, something better than the prosthetics available. I regret that, although I did later make an artificial hand and take out a provisional patent for controlling it by brain thought waves, I never made an artificial leg for him. I met up with him much later in Malaysia, and he was happy and content with a family and a successful engineering career, running around on the much-improved legs then available. What it did for me was turn my attention to more than just pattern recognition and artificial vision. So, it was no wonder that just a few years later I said, 'I can do that,' when Sandra Wills came to me from Elizabeth College asking if I could build a robot turtle.

Actually, my first attempt at a whole physical robot was a little before that. I had been repairing jukeboxes at the Automatic Music Company in Hobart. I removed a flat wooden box that served as a woofer horn from the top of an ancient jukebox to use as the base of a robot. My schoolmate David Briers the school's most athletic performer, who taught me to wrestle and to jump fearlessly off high house rooftops, who became a fitter and turner, made the mechanical propulsion pieces for me. Apart from running around with a battery and some motors, though, it never progressed. Too heavy and bulky and impractical. My idea of what constituted a robot was naïve back then, but it taught me what not to do for what became the Tasman Turtle design.

Originally conceived of for teaching educational classes using the computer language Logo, the Tasman Turtle, when it was completed and commercialised, served to fill an unexpected niche for much more than what it was designed for. It quickly came to serve as a model for sophisticated hobby and large toy robots like Elami and Omnibot for companies like Keystone and Tomy, but more importantly it opened up doors with the likes of Radio Shack and Commodore and General Electric and, in turn, networking opportunities with names like Nolan Bushnell, Jack Tramiel, Bernie Appel, Rodney Brooks, Hans Moravec, Isaac Asimov, and the Father of Robotics, Joe Engelberger, who co-invented and developed the robot arm with George Devol, forming the first robotics company Unimation. There was clearly a need for a useful, practical robot chassis that could serve as the basis of anything an aspiring roboticist wanted. My company and my product and my name became solidified in the international robotics sector.

And so it begins. I relish the spotlight. I discover there was no other enterprise like mine, anywhere. My company, starting as Aero Electronics, then Branch and Associated, then Flexible Systems and finally Denning Branch International, becomes the prototype robotics company for the commercial side of the industry, the Tasman Turtle the prototype general purpose robot. I become the industry's prototype spokesperson. I'm away.

Those external projects with such important names and companies required me, as the inventor and proponent of my company's technology, to lead their projects hands-on, either completing them if they had stalled, or starting them if they were just dreams. So, this was the beginning of a globe-trotting career, transferring my technology and skills, under licence or paid, to exciting companies in exciting places with exciting people. This was a fast-paced time, quick contracts, rapid engagements, loose arrangements because of the vague project definitions and need for an indefinable solution, a period for me that could be interpreted as outsourcing, or equally as an inhouse consultant under secondment.

For 20 years I was part of that scene, but in a funny way I was the odd man out. Apart from a couple of later companies like Cybermotion, the robotics activity in those early years was mainly centred on academic research with fantastic but one-off research vehicles like Han's Pluto. Those robots had no commercial aspirations. Instead I was out there making and selling toy, hobby, educational, industrial, research and domestic robots. Devices that performed a function that people were willing to pay money for. It was more about marketing than technology.

Unlike in Australia where there is little scientific research inside companies, the preference being for universities, I had moved into that uniquely American rarefied zone of commerce driven by internal corporate R&D and IP. I was not running a university research lab, I was running my private lab with a bunch of PhDs working with me. That close association with scientists meant there was always a vestigial longing within me to be part of academia. I was making scientific discoveries that were trade secrets that I could not publish or divulge.

This networking sometimes led to other problems. John Holland at Cybermotion had become a friend. We had stayed at each other's residences when we visited each other, me in New York, John in Roanoke, Virginia. When I took over Denning, it turned out that Denning had used a three-wheeled omnidirectional motion system similar to John's robot Kludge. It was nothing to do with me, but he interpreted my involvement with Denning as dishonour and our friendship ended.

Until I returned to Hobart from Texas to develop Blinker, apart from some simplistic navigation and mapping through touch and feel, by bumping into a wall or object and sensing it with a touch sensor, or using sonar to detect objects nearby, all of my robots navigated in an unreliable and inaccurate mode called dead reckoning. Basically, dead reckoning is blindly counting an estimate of how far had been travelled and in what direction from wheel encoders or by measuring how much time had transpired. A guessing game. It's exactly like those early sailors and explorers trying to estimate longitude before John Harrison's amazing clock allowed them to compute it from time differences. Harrison, a self-made inventor, had to fight the academic establishment to be recognised for his achievements. Shaft encoders to count wheel rotations increase accuracy but errors can occur because of slipping and friction. True autonomous navigation, necessary for effective performance of any useful task by a robot, requires a map and an ability to know at any time where the robot is within the map. Maps can be entered in advance, like a floor plan, but more useful and more complex are technologies where a robot operates in an unknown and unstructured environment, building the map itself, through vision, sonar, radar or other sensors more advanced than bumpers or wheel encoders.

Creating such a technology became my obsession.

Taking this to a more pragmatic level is an even greater challenge when the operator of the robot is a non-technical person who cannot enter a programmed map, or do anything other than turn it on or off, to operate the robot in its environment. A quintessential example of that is floor cleaning. Few people vacuuming a floor know or care about motors and fans and vacuums and filters. A vacuum cleaner robot is a vacuum cleaner. An appliance. The average person using a robot vacuum cleaner would know nothing of robotics. This led to my interest in the ergonomics of design, the skill to create complex and advanced automation systems at a low price and with good human machine interfaces. These were huge steps for someone starting off as a run-of-the-mill techno-nerd. It is probably lack of attention to these things that prevented, still prevent, many promising inventors from commercial success.

Outside of the markets I already dominated – education, hobby, research – it was a struggle to find new near-term markets. There were a few smallish ones, like disaster robots and bomb disposal robots, but others required the science fiction of humanoids. One potential market was the one just mentioned, domestic cleaning. I always believed that there would be a market for a robot vacuum cleaner, but one of the tenets that I established of turning around companies was to conduct independent market research, in case you are fooling yourself. I contracted well-known Australian research firm Roy Morgan, which discovered in a national survey that there was indeed a huge multibillion dollar market for such an appliance. That was not enough, though; I needed to verify it with a duplicate survey. I hired my sister Olannah, and we conducted our own in-house phone market research using the same Roy Morgan survey questions, statewide, which on extrapolation generated the same results. The exercise taught us how to do such unbiased market research for other products. In 1987 the initial global market size for a practical domestic robot vacuum cleaner was 12 billion dollars.

The survey research not only uncovered the massive size of the robotic domestic vacuum cleaning market but exposed some key product specifications about retail price and product functionality that were necessary if the market potential was to be reached. If these were not met, the product sales deteriorated dramatically. From this invaluable pricing information, I discovered the wonderfully useful Optimum Pricing Curve, and calculated the best price to make the most profit from sales of such a device. That was another revelation: that it was possible to determine without trial and error the introductory price of a new category of product. It was a bit like my amazement in early high school lessons, learning about Newton's mechanics, and discovering that it is possible to predict the result of momentum and forces, like where a missile will land.

The first project to create a robot vacuum cleaner came to me out of the blue; the Florbot robot for General Electric Plastics (more about that in Chapter 13). GEP did not manufacture vacuum cleaners, of course, but they supplied most of the plastic to those manufacturers who did make them. The industry was flat, no growth, because as a mature industry, there was no innovation. All industry sales were essentially what are called maintenance sales; a customer buys a new one when their old one wears out or breaks down. Marketers get over this problem by making new designs, making the old ones obsolete. We see it with cars all the time. A perfectly good car starts to look dated, and psychologically we are teased into updating it. GEP decided to stimulate plastic sales by doing the innovation work for their customers. They saw a robot as a new category of product that could boost their plastics sales.

The second opportunity came to me unsolicited, too, from Moulinex in France. Like many of the robotic projects that came to us from companies noticing the Tasman Turtle, Moulinex had started a robot vacuum cleaner project, which proved what everyone else had discovered: that it was unexpectedly difficult. There were few options for them to reach out to for assistance, so, of course, they discovered me. We built d'Entrecasteaux for them.

Following those projects, travelling the world, I presented the opportunity to all the major global players in the vacuum cleaner market, together with the market research. Several of them took up the project, but without me. When I disclose sensitive inside information I reserve something for my own security. That was the approach that worked with GE Plastics. For the vacuum cleaner companies I disclosed the market research, but not the functionality or pricing research. All of them applied misguided pricing strategies, ignoring independent market research available from me; these companies set their own pricing, typically exorbitantly high, resulting notably in one case, in zero sales. Mistakenly, those companies believed that the demand for such an amazing appliance that cleaned the floor autonomously must demand a high price. They ignored the 'elastic market' rule that caps the price for certain product categories. They knew marketing 101 but not marketing 202.

I was visiting Rodney Brooks at MIT in the mid-1990s, maybe 1997, and showed him some video of our autonomous robot navigation projects. He clearly had never been aware of our work and he was beside himself with excitement. Rodney, an Australian, at that time head of the robotics activities at MIT (who together with Hans Moravec who was at CMU) was one of the most preeminent robot researchers anywhere. These were the scientists carrying on from the work of William Grey Walter. Having such a reaction from Rodney was alarming, unexpected. He grabbed me in his fervour and said, 'Come with me.' We walked from Cambridge down a few blocks to his private company iRobot, run by his cofounders Colin Angle and Helen Greiner.

Back then they were a tiny backwater robotics company with a handful of employees, surviving off small federal government and defence research contracts, but clearly in a strong market position because of the MIT connection. Like all of my competitors I was very much informed about them and had briefly met Colin and Helen on a previous occasion. Once we were in their office Rodney disappeared with my videos to meet with Colin and Helen, then came back saying he wanted me to join iRobot with them and had spoken to his cofounders. They had stayed in their offices and I did not see them. My memory is that Colin was in agreement, but Helen could not be convinced. Without a consensus it did not happen. iRobot subsequently released their Roomba robot floor vacuum cleaner. They stuck to the pricing rules determined by the Roy Morgan research. Such a price limit proved difficult for even iRobot to install full functionality to their robot, so the first one was a device with much lesser performance than we had in our prototypes, but with their name and having raised funding through an IPO, they have subsequently sold and are selling billions of dollars worth of robot vacuum cleaners, as my market research predicted.

At this time I was very interested in why there was such a tight price constraint on such a complex and useful advanced technology. I considered other technologies such as medical breakthroughs. If a cure for cancer were to be discovered, the inventor could demand whatever price they wanted, and the demand would be there. If there was a discovery of a manyfold increase in telecommunications bandwidth, the price would not be an issue because of the demand. Why, with a smart, autonomous robot, did the same rule not apply? It turns out that in the case of those other examples, there is no alternative, no competitor, whereas with robots there is a powerful alternative, a competitor. The competitor is the human that the robot is attempting to replace. A human doing vacuum cleaning, for instance, is an unskilled worker with a set maximum hourly wage, which is low. Even an unskilled worker, though, is a very effective cleaner, far superior to any robot vacuum cleaner, so the performance of the competitor is very high. Unless a robot vacuum cleaner is as cheap as human labour and can clean as effectively, then there is limited market demand and the customer will hire a human cleaner instead of buying a robot. That was why the other companies that had good machines, better than Roomba, but prices ten times higher, could not sell them. A human can prepare the room for the vacuum cleaner, move furniture around, does not get stuck in corners, can see what still needs extra cleaning, and do all those ancillary tasks critical to vacuum cleaning. Once a robot achieves such functionality the market size will be incalculable.

When there is a competitor like that, the product has to be differentiated. Pricing is one differentiator, but there are better ones. Robots do not steal, sleep on the job, demand additional wages, can work day or night even with the lights off. So, once the price and functionality has been matched, the robot has other benefits.

I wrote all of this up in a paper called 'The 15 Year Cycle of Robotics', published it, then presented it at lectures, seminars and conferences. I was accountable for about 60 per cent of the global robotics industry at that time; however, in 1997 I subsequently exited the industry I had pioneered in 1979, forever. It was one of the hardest decisions in my life, but I'd had to admit to myself that there was no market for robots. Yes, they were always topical, of course, and always exciting, there's always research being done on them and they're always in movies and novels, but decades later, apart from the Roombas wandering around or shelved in closets, no one owns a robot.

The next stage in my life, my Middle Ages, was like driving down a rutted road, the car going where the ruts take it, not where one plans to go.

CHAPTER 14

WATERLOGGED

The first family dog I can remember was my grandfather's dog Nokie. I think that is how it would be spelled. Pa Branch was a short, gruff man, jowly-faced, thin hair. In old age he wore trousers too large for him, with the excess material tucked over a belt around the waist, the belt sometimes a piece of cord. Like men in all the old sepia pictures of his generation, he always wore a battered felt trilby and suit jacket. He was always old to my child's eyes. We were all wary of him, including Nan, although I rarely recall seeing him. Straight from work to the pub, then home to his own bedroom. I have few images of him. I can only faintly recall his dog Nokie, who lived outside. Towards the end of his life, when I was in my mid-teens, Pa visited us and out of character he came into the garage to talk with me where I was toying with parts and panels from my dad's wrecked car. He looked at what I was doing and explained that panel beating is done by hammering the metal over a sandbag. It was a kind, useful, technical tutoring and I like to remember him through this one interaction. During a recent visit to my brother Toni, who was in his final stages of terminal cancer, and who had refused to read the draft of this memoir, he related an identical story, without any knowledge of this interaction of mine with my grandfather. As a hard labourer in an early job, digging foundations for the new university buildings, Toni came home with bleeding hands. Home for Toni was all over the place, but at this time it was with Nan and Pa. Pa, as familiar with hard, physical labour as anyone, took him aside and said the best remedy was to rub suet into his hands. Toni told me how Pa, delicately because of the pain, held his hands and with the softest touch massaged the raw fat into his bleeding callouses.

This was also the first time I realised that teachers don't know everything. At primary school, once a week we had to write an essay, a story, about anything. We all struggled to fill a page; a foolscap page back then. I decided to write about Nokie and I asked the teacher how to spell it. She asked me to repeat the word, which I did, and she asked me what it meant. I said it was the name of a dog. She said she did not know, but try N-o-k-i-e. Try? She didn't know? Nan used to have a photo of Nokie balanced in the front carry basket of my Uncle David's bike. It had been published in the newspaper, and now Uncle David's son Peter, my cousin, has it. Only after my Uncle David recently died, and the photo and the story revived at a family meeting, did I discover that the dog's actual name was Pinocchio, because it was a Jack Russell, with a long nose. Nokie was its nickname. I never knew that.

We always had dogs. Most of them got distemper and died. Stockholm Tar on their nose did nothing, like many home remedies. I cannot recall any of them getting vaccinated or going to a vet for anything. So, some were with us a short time and I cannot remember their names, but I can see them in my mind. I cared for them and hurt for them. Our main family dog was Ricky, a beautiful, friendly, golden collie-shepherd cross. He was poisoned by bait from a neighbour when I was about 11 years old. I can see my mum, devastated and angry because she knew who the culprit was, but it would be futile trying to prove anything.

When I was married to Gay, living on the hobby farm at Lonnavale, we had two dogs. A scruffy, friendly, grey Smithfield called Sebastian, which was a name Gay always wanted to call a dog, and the other a slim black bitser called Chien, the French word for dog.

My dog, though, the one I think of as mine, was Kari. Mim and I had been together for a bit under a decade when she bought her first dog, a blue merle collie rough, like Lassie, but mottled grey and white and black. That's what merle means. It can also be a name, like in Merle Haggard, one of my favourite country singers. In that sense it comes from the word for blackbird. Kari means 'smoke' in a South Australian Aboriginal language, which was the name she chose for him. It was her dog, but to me he was mine. Later she got more dogs, and I loved them all, and miss them all, cried when they died, but I loved Kari as one only loves their first love. Puppy love, literally. As Mim's dog numbers grew, we purchased a rural property at Wattle Hill, a half hour out of Hobart. It was a beautiful, secluded farmhouse with a large yard of kennels and runs and fences and puppy nurseries and dog grooming rooms and training paraphernalia, where loud noises, canine or human or machine, were inaudible to any distant neighbour. Where life and its activities were invisible, hidden by trees and distance. The Iron Creek borders the property; it's dry in summer, a raging, thrilling torrent in winter, which is a temptation for children who visit. And tempting for dogs too.

This is a story not of a man's dog, but of a dog's man. It is a story of how to suffer from hypothermia while being ignored by all the pet-fretting friends around you.

While Kari was cuddled up in cosy woolly rugs, heated by a huge open fire and treated to luscious, smooth, blood-warming red bourbon, I was ignored, left to shiver uncontrollably and suffer in absentia, never once treated as the heroic human I personally believe I had just been.

Kari always let us know when someone visited. Living in the country away from the fast roads and highways, wrapped up in our secret world of loud music and classic movies, it is not always easy to hear a visitor coming along our winding path, with the babbling brook close by and the ark of animals around us. Anyone thinking that the country is quiet compared to the city is horribly mistaken. Only here can you hear the grunts of animals of all species. Trees creaking and cracking, the wind rustling the leaves. Young kids, when asked what causes the wind, often think it is the trees. There were noises galore on that night, with its crashing after-effects of the recent storm and the torrential high waters of the river, now almost up to the roadway. Fervent squalls blew the famous Tasmanian, cold, three-day south-easter across our home. Inside were two roaring fires.

Kari was outside. We were comfortable inside listening earnestly to the storm when some close friends arrived. So excited were we that it was some time before we realised that, out of character, Kari had not let us know these people had dared trespass. It does not matter that they were close friends, frequent visitors, and well known to Kari – he would not let anyone dare come near without a warning. Perhaps he knew what we got up to inside and wanted to spare us the embarrassment of being found out. I have to admit that some of those movies are really old and some of the music is really off, and we quickly shut everything down when someone comes. It saves them the embarrassment of having to say that it's okay, leave it on.

After a short while, Mim realised that there had been no warning barks from Kari. Not really alarmed but still a little apprehensive because it was different, she went out to call his name. Perhaps he really was simply used to these dear old friends by now and just did not bother to tell us when they were coming up the drive ten minutes or so earlier. He was quite renowned for not bothering with some things, like a child: enthusiastic to do what he wants, but slow to do anything not high on his list. But a dog not taking the opportunity to bark? Now really! No response came from Kari as Mim called his name, and then she really did become concerned, calling his name out everywhere as loudly as she could. We were inside deciding if the dated sounds of Jimmie Rodgers' music from our CD collection playing on the compact disc player should be politely turned off, and unaware of this searching. Remember CDs? Remember record collections? In fact, I remember now that I had just received an important incoming international telephone call, and was in the opening of intense business negotiations with a customer I had never spoken with before.

Suddenly Mim rushed into the house demanding that I come quickly. 'Come quickly, come quickly, I have a problem with Kari!'

I looked at the telephone in my hand, looked at Mim, then looked back at the phone, like Indiana Jones deciding between helping his desperate friends or finishing reeling in the fish he had just caught on his rod. I said to the person on the other end of the phone, 'I have to go,' promptly hanging up on him with no further explanation and not waiting for a reply. To this day I do not remember who that person was, and he has never called me back.

We all rushed out of the house, caught off guard, not knowing what the problem was in those immediate few seconds and, of course, fearing the worst. We followed Mim and raced to the edge of the river, pitch black in the storm-cloud-covered night, slipping in the wet mud and tripping over fallen branches, finally reaching the edge of the riverbank. Churning water boiled 6 to 8 feet below the ground level, and below we could see poor Kari clinging for dear life to the edge. Just a little further downstream, the river widened and came closer to the lower siding in slower running and shallower pools. Kari was clinging for life at the deepest part of the river in our vicinity and the highest side bank. If Kari had simply let go and floated downstream about 50 feet, he would have been able to struggle ashore and save himself, but in the panic, his instinctive reaction was to cling to the bank, believing that to let go would mean certain death, taken over by the enormous pressures of the water and washed out to sea. That is, if dogs think about things that way.

After no more than seconds of trying to reach down to pull Kari up over the edge and back to the ground, and discovering that it was futile, I jumped in. My shoes went off beforehand, but other than that I was fully clad.

Now this part of the river is usually about waist deep in the height of summer, so no one was more surprised than me when I quickly went completely out of sight. Underwater and with the bottom nowhere near my feet, I bobbed cork-like up and down a few times then started to wash out to sea, at least in my mind, because humans do think that way. Somehow, I managed to struggle from the centre of the river to the bank and I grasped out for Kari. Kari had no strength left, and at the moment of touch he let go and all of a sudden there I was again drowning with the weight of a large, waterlogged dog trying to cling to me as if I were an island, and surely moving away from the safety of the edge again.

All I remember, and I remember it vividly, was the complete trust and confidence the dog had in me, its master and friend, that the minute I was there in the water with him, I would save him. To this day I can't think of this episode without recalling that dog's love and dependence on me at that critical moment radiating through everything else that was going on. It brings tears to my eyes. Everyone else was on the bank, and I think there was lots of noise, but I was in that small world drowned out by the sound of powerful freezing-cold water, reacting to the instincts of survival.

Eventually I managed to get to the edge and grab onto the nearest bough. The next dilemma was how to get Kari, then myself, up out of the water and onto dry land. Well, slippery and wet and muddy dry land. I pushed him into the air in the general direction of the humans on shore, and they pulled and the dog made it. They carried Kari back towards the house. With so much consternation over Kari I was left to get myself out, using overhanging branches from the wild bushes and trees.

After retrieving my shoes, I finally caught up with them and we all raced into the house, dripping river water all over the floors, then nesting eventually beside the roaring log fire. We all became instant experts in treating hypothermia. Kari was wrapped in a blanket, hot water bottles were fetched, some alcohol was administered, to Kari, not the humans, and he was kept as close to the fire as was safe. He was very weak. Only much later did we realise how close a call it was. Kari had behavioural disturbances for almost a year, and took longer than that to regain his normal weight. This did not deter him from getting into the river again and from time to time getting entangled in the bushes along the edge, but luckily no more dramas like that one night.

I was still standing in my muddy clothes and soaked shoes which I had put on before racing to catch up to the others, not aware of my own shivering. Eventually the others realised that there were two hypothermia sufferers, and in my case it was a hot bathtub that brought my temperature around. No hot water bottles, no medicinals, no room next to the fire and no blankets. I even filled my own bath, shivering so hard that I almost could not turn the taps on.

A little while after returning to the house, we discovered blood everywhere. The obvious choice was to search Kari, but there were no cuts. It turned out to be a huge gash on my hand, and the next day we found that the one branch that I had grabbed to get a firm purchase when hoisting Kari and myself out of the river was a box thorn, with sharp 2-inch spikes that are lethal weapons. I had grabbed straight onto a thorn, but had not been aware of it even after we found the blood in the house. Adrenaline and endorphins are wonderful drugs.

In the few moments that I was in that river I lost my watch. An inexpensive but precious gift from Mim that I had had for more than ten years. I searched for days for it and even years later I still went down to the river when it was low in summer hoping to find it jammed under a rock or sunken branch. At one time I borrowed a snorkelling mask and swam around for hours with my face under the water looking for that watch. That time I had to come back inside and run a hot bath again because I had been in the water for too long and had lost all my body heat. Déjà vu.

The visit by our friends turned out to be the most fortuitous one ever. Kari had escaped his kennel run that night and must have been in the water for ages before they came to visit. Only his absent bark saved him. I asked Mim how she knew he was in the river, because our farm has innumerable exciting places for a dog to explore, and she said that when she called his name she could just hear this weak whimper from the water, only just loud enough to be heard over the noise of the water. The poor dog had not enough energy left to bark. If all of these events had not happened as they did, Kari would not have survived, and the world would be a poorer place.

On the night, though, we had a mixture of emotions; we were relieved to have found and saved him and angry because of the anguish he caused us. It did not occur to me as I was trying to get myself ashore, after using all my energy to get safely away from the fast water in the centre of the river, then hoisting Kari into the air, that I could have done what I thought Kari should have done: simply let go and let the water carry me to the shallow and easy access section a little downstream. Instincts are hard to ignore;

there was no way that I was going to let the water take over me, even for a few seconds and a few feet.

And after it all, Jimmie Rodgers was still singing 'Blue Yodel' in repeat mode on the CD, with no one bothering to think about whether it would be more prudent to actually turn it off.

Years later when Kari died, I resisted letting him go. For two days I kept him on a blanket on the floor in the guestroom and I would go in to visit him, lie down beside him. Then finally I dug a grave on the property where he roamed freely, near the river that he frequented, and buried him in the same blanket.

CHAPTER 15

BILLY BOB'S TEXAS

Every Saturday I would lead my brother and sisters like ducklings across the city to the matinee movie. I had five siblings, and my memory feels like the train consisted of six kids, but Carol had not been born and Dianna would have been too young, and I would have been less than ten years old, because just before I turned ten we moved to a different home away from the city centre where all the movie theatres were located. So, there were probably only three of us. It feels like more in my memory. And I feel like it was a long-standing tradition. But memory is unreliable and it might have only been a few times. From 28 Regent Street in the suburb of Sandy Bay (not far from where Errol Flynn lived) to the CBD was perhaps a safe, leisurely 2 miles, pre-metric days. It is still 2 miles post-metric, but decades have made the trip dangerous. Once quiet suburban roads are now noisy thoroughfares of people and traffic, and young kids are not allowed to risk walking by themselves in the street anymore, so, they are growing up with no road skills. Makes no sense to me; like a pet dog let loose for the first time into a world it knows nothing of.

For kids it was, I think, a shilling, called a bob pre-metric, per child to get in. Could have been sixpence, a zac, or thrippence, which was a trey, to see the double feature separated by intermission, with a newsreel to start and a cartoon. Value for money. We went regardless of what was playing. It would be years before I heard slang terms like nickel, dime or quarter.

At Waddamana on a Wednesday night, for kids it was thrippence to watch cheap older movies recycled by the Hydro to entertain the workers. In the dark I'd walk through the couple of streets around houses, across a small field, bypass a shallow creek, cut my hand on a shaft of blade grass trying to balance in the darkness, then stream in with the other patrons as we reached the community hall, with its forms and stools. The main movie I remember best is one where they had puppetry or stop-motion T-Rexes and fur-covered cavemen in the same scenes, so I was absolutely sure for years that humans and dinosaurs lived at the same time.

Another movie I saw was *The World of Suzie Wong*. No such thing as age ratings in Waddamana; everyone went to all the weekly movies. I would have been about ten years old then. I remembered it so vividly that years later I was ecstatic when I lived and worked in WanChai, still a redlight district in Hong Kong, still endearingly called Suzie Wong District even now.

Promoted heavily in 1959, *Ben-Hur* starred the square-jawed Charlton Heston, a great stage name for a Scot whose name was really John Charles Carter. On this Saturday, off we went again with our shilling pieces to see the week's movie. We usually, but not always, had an extra zac or bob if we were lucky, or if Dad was generous or had the money, to buy a packet of lollies. Never heard the words 'sweets' or 'candy' back then or, heaven forbid, the uppity word 'confectionary'. 'Sweets' maybe, but that was usually reserved for dessert. Our favourite lollies were Milk Shakes, white vanilla-flavoured toffees. They were a bit more pricey, because they were rock hard and could be sucked forever. As soon as we arrived outside the Avalon Theatre in Melville Street it was clear that something was different. The lines were longer, thanks to the power of promotion. There were more adults than kids, which was unusual, and it was taking longer to be admitted via the box office, which are located inside the theatre lobbies in Australia. Never heard the word lobby back then. We were in line, slowly creeping along, filled with anticipation and excitement. Unknown to us, this was not a B Western with cowboy actors like Randolph Scott or my personal hero, Audie Murphy.

Eventually, we were at the counter and I handed over the money. We were regulars and I recognised the ticketmaster, but there was a problem. He looked at me and at my money then at the crowd, clearly embarrassed. It took a while, but there had been some sort of communication between him and someone off to the side, and we received our tickets. As I turned around, I saw the manager of the theatre, standing arms folded, nodding his consent to the person in the box office. We had no idea that the price for this special feature was not the usual shilling but 2 shillings and sixpence – a crown – but we urchins all got in cheap. The power of promotion. Later, the theatre manager, whose name I wish I could remember because it was such a generous thing to have done for a bunch of poor urchins and I will never forget it, managed the local Allan's music store from where I purchased my first guitar, a Fender Coronado, a sunburst-coloured single pick-up semi-acoustic. He was a person around town.

Dad renovated every house we lived in. At times he ripped out walls and flooring and added rooms, bringing outside toilets inside. In the remodelling rubble of the house we lived in before the one from where we trekked across to the movies, at 41 Goulbourn Street, right in the heart of the city, I found a holey dollar. In the very early days of the New South Wales settlement in Australia, there was no coin mint and no local currency; in fact, issuance of currency was forbidden in English colonies. One type of coin that had been floating around was the Spanish dollar, the famous Piece of Eight. They were ubiquitous, and were valued at eight reales, hence the name. There was still a shortage of coinage, though. Sometimes these Spanish coins were cut into eight little bits like a pizza, but the tiny pie slices were easily lost. Australians had another device: we punched the centre out, creating two coins still recognisably round in shape. The large outer annulus with the hole in the middle was called a holey dollar, a fun play on the religious meaning of holy, perhaps because it resembled a halo. The centre, like a punched-out donut was called a dump, another Australian vernacular for something quite different these days. As a four or five-year-old I knew what I had found was money, just an old coin that I could buy something with. Happy-go-lucky Mr and Mrs Smith, Ma and Pa Kettle lookalikes, had the corner grocery store two doors up and I remember walking up and handing the coin over to one of their two sons, Peter Smith, who was maybe about 13 years old, asking what I could buy with it. What do you want? he asked. Of course, I wanted lollies. I should have known it was valuable because he opened the lids to the glass jars of lollies across the counter and said to choose whatever I wanted – something that never usually happened. Today that coin would be worth around $40,000.

Ben-Hur was a spectacle. Unsurprisingly, the expression in Australia for anything humungous became 'Bigger than Ben-Hur'.

I love country music and American honky-tonks, and the Ben-Hur of honky-tonks is a converted stockyard in Fort Worth, Texas.

Fort Worth in Texas used to be the rail head for freight trains to the markets in Chicago. Cowboys would drive the cattle across Texas to the far north to Fort Worth, then corral the stock there to be rested and fattened up after the long trek before being loaded onto the trains for the final trip to the markets. Fort Worth's nickname is Cowtown. Texas is a big state and has lots of cattle.

The cattle yards in Fort Worth cover 100,000 square feet. Trains are everywhere nowadays, so this is no longer the holding ground at the end of the trail, Randolph and Audie eat your hearts out. But Texans waste nothing, and the real estate is valuable. What else does Texas have lots of? Cars, pick-ups, with gun racks across the back windows ... but where there are cowboys there is cowboy music, country music, and in 1981 the Fort Worth stockyards were converted by Billy Bob Barnett into the world's largest honky-tonk. Billy Bob was a famous footballer and graduate of Texas A&M University in College Station, where I later delivered guest lectures on robotics.

Everything Texas does is bigger than anywhere else, and nothing prepared me for Texas the first time I arrived there. I'd been to San Francisco and Boston, but on my flight into the airport in 1983, squinting into the too-hot, eye-burning bright sunshine, I could see miles of multilane highways, lined all the way with more commerce that I had ever seen before. Hotels, stores, service stations, restaurants, fast food outlets, shopping malls, theme parks, all swamped under flashing neon signs. And it was still early morning. From 1983, my partner Mim and I lived in Dallas, Texas, which is a hop, skip and jump, Texas-style, across from Fort Worth. Twin cities. I worked for Commodore Computers, the world's largest PC company at the time, as head of robot technology in their research centre in North Dallas. The huge airport is placed midway between the twin cities. My route was from the airport straight to north Dallas, immediately to the office of Commodore Computers, where I was expected to start work, despite that I had just completed a 30-hour international flight from Australia.

I met the most enthusiastic team I have ever worked with: Richard, Tom, Ed and Ben, who was sporting a hand cast because his excitement at a technical breakthrough earlier resulted in him hitting his fist too hard against a solid surface. It was my introduction, not to the US but to Texas, a different place. Thirty years earlier, Leslie Poles Harley's character in *The Go-Between* may have said, 'The past is a foreign country,' but Texas is definitely a foreign country to the rest of the USA. They do things very differently there, including that men still shoot each other. I loved the country music and the American superstars, watching *Hee Haw* on TV with superstars Roy Clark and Buck Owens, the Grand Ole Opry broadcast from Nashville on million-watt radio stations, Jimmy Dean selling sausages instead of singing about Big Bad John. Everything everywhere was bigger and better than the rest of the USA.

More than anything, I was sad that I had progressed to be so close to the music and musicians that were like folklore to me and my family but my family were missing out. They would have loved it all. They lived for music and all it entailed: parties, dancing, drinking, revelry. But while I was there in Texas, they were dying: my Aunt Phil then my dad in quick succession while I was in Dallas. Aunts Doreen and Molly and uncles Lionel and Cedric all had died a short time earlier, as had my grandparents.

Only Uncle David was still around. We all shared our music, visiting each other with records of our latest discovery to play for each other. I was unable to share this experience with them. 'You cannot stop yourself from thinking about all the others who never managed to get as far…' as Paul Auster says in his *Winter Journal*.

Dad had visibly deteriorated before I left, the culmination of a lifetime of poor lifestyle choices. Before I left, I gave him the telephone number of my new office in the USA and an explanation of how to make a direct call, including what times, but I was nevertheless surprised as I sat in my sparse office on a weekend, not a time I suggested Dad choose if he called me, to pick up the phone and hear Dad's voice. He was just as surprised that it worked, and that is almost all he said, 'It worked.' They were the last words I ever heard from him. I think he knew something was wrong with him. A few months later my sister Olannah called to advise that he was in hospital, diagnosed with an incurable brain tumour, and might not come out. I raced home to Tasmania, and held his hand the next day as he passed away after falling into a coma after surgery. I was screaming at the top of my voice in the public hospital ward.

'It is me, Allan, I am home, Dad. Dad! Dad!' Hoping he would hear me somehow.

For many years I would wake up from a restless sleep fantasising that he was still alive, with tear-saturated face and pillows and the realisation it was only a cruel dream. It wasn't until years later when I woke up like this one night, and wrote my final letter to him under the dim single lightbulb in the kitchen, that the dreams stopped and I felt I had said my goodbyes.

So, he never got the opportunity to join me, to share in the impossible-to-comprehend opportunity for a small-town Tasmanian to actually be in the midst of the Texas country music scene.

As soon as I discovered Billy Bob's, it became a regular haunt. Forty minutes from my condo in Dallas, past the huge Six Flags Over Texas water theme park, to 20 acres of unpaved car park, an indoor rodeo ring, seven dance floors, three stages, shops and restaurants and galleries and souvenirs. There we saw 'King of the Road' singer Roger Miller, who had just revived his career with the successful play *Big River*, the Osmonds singing country instead of pop, local Texan Claude Gray, and the yodelling Slim Whitman. Years earlier in Hobart I saw Slim Whitman in his spell-binding concert in our primitive city hall. He played all the songs I had heard since an infant through my family's constant partying and singing and dancing and drinking. At Billy Bob's I walked up to the stage and put a note on a napkin on the stage in front of him; it said that I had seen him before in Hobart and requested that he play 'Indian Love Call'. He saw me, but ignored the note for a long time, tipping it around a bit with the toes of his alligator leather shoes. Then, he picked it up and read it. Still no reaction. I was disappointed, until he finally stopped between songs and said over the microphone: 'In the audience tonight we have Allan Branch, all the way from Australia. Allan saw our show in Tasmania a few year ago, and loved it so much he has come all the way here tonight to see it again.' Talk about gall! I had to laugh, but I was thrilled.

When the spoof sci-fi movie *Mars Attacks!* was released, it was perplexing to watch the finale where the Martians' brains explode accompanied by the yodelling in 'Indian Love Call'. It's funny but it blows my mind too.

A penny arcade inside Billy Bob's had one of those punching bags that measures your straight right punching power. I am reasonably strong and was having a good punch when a thin, scrawny, young Hispanic cowboy came up, hit the bag and doubled my score. I was amazed. We got talking and he showed me his trick. He had a heavy metal bolt in his fist, so that the momentum of his punch was like that of a sledgehammer. He gave me a try with the bolt and I beat his score. It was satisfying and the cowboy was unfazed and pleased for me, that I was now punching above my weight.

As we left and walked to the car in the carpark, we saw a crowd of people fighting and heard a crash. Not unusual in Texas. Nor in my family for that matter. When we got to the car we discovered that someone in the fight had thrown a brick and it had smashed my car driver's side window. It was easy in Dallas to find a car wrecker's, buy a replacement window, then, with the tools from the Commodore workshop, repair the damage. The mechanical skills learned from my uncles came to the fore yet again.

CHAPTER 16

KANNA LENNA

In 1876 the world's first hydroelectric turbine was established in England by William Armstrong to light his house at Cragside in Northumberland. Not as trivial as it sounds, Cragside being the size of a small village. It was only 40 years from the actual invention of the generator by Michael Faraday, a primitive device and really only a demonstration of the induction principle – how a magnetic field from a wire with electricity here will induce an electric current in a close by wire there. Just another 40 years later, the Tasmanian state government's hydroelectric power station was operating at Waddamana, a village hidden in a deep valley among craggy mountains and silver lakes in the literal geographical; centre of the state. Visionary is the only way to describe the plans and aspirations of the team that built such a revolutionary, pun intended, infrastructure. To me it is akin to the dream in 1962 of putting a man on the moon by 1970. Daring, idealistic, risky.

The hum of massive vertical Pelton wheels spinning 50 times a second day and night from anywhere in the Waddamana village are among my earliest auditory memories. That and standing on the green metal viewing gantry high on the main wall of the incongruous art deco power station building, looking down on the rows of energetic turbines like terracotta sentries, or as cousin Greg suggested, terracotta snails, the noise preventing any conversation.

Waddamana was not the first electrical generator in the state. Amazingly, the first was in 1895, less than a decade after Cragside. That is almost impossible to put into perspective for a tiny state with a tiny population at the tail end of the world; this kind of accomplishment was more expected in the laboratories of someone like Edison in New Jersey. But there it is, Duck Reach Power Station, the first hydroelectric generator in the Southern Hemisphere, powering the arc lamps of the sleepy hollow called Launceston, the work of the Launceston city civil engineer with the unlikely name of Charles St John David. But it was, like they say, in the air; there were other small private generators; one in Queenstown, for instance, when this now almost abandoned centre for Mt Lyell copper mines was once one of the state's largest towns. Mining generates money, generates commerce, generates invention. Australia's first stock exchange, for instance, was in the remote Queensland mining town of Charters Towers. In Tasmania, it seems, mining generated generators.

Charles Wesley was my science teacher and home class master for my first year at high school. He was very English, from England, and I remember learning about soft and hard water, despite Tasmania having no hard water, not understanding the relevance. And I remember his chemistry lessons, enthralling little bombs and explosions and smoke and smells. I learned that metals like zinc and aluminium required lots of electricity to electrolyse the ore and extract the minerals. Wesley was gruff, a prototype for Trevor Howard's Captain Bligh and just as intimidating. So competent and knowledgeable about things that, like Bligh, he surely would have been able to navigate a dingy to shore from anywhere in the ocean. One evening's excursion was to the school yards to identify all the stars, as if each were his favourite pupils, and I can still dole out the members of upside-down Orion, hunting on his head, and remember that there are seven sisters in Pleiades. I'm a Gemini and I learned to recognise Castor and Pollux. Like Bligh, Wesley would have navigated by the stars. I never had a better teacher. He secretly gave me lab glassware and chemicals to encourage my experimenting hobby at home, to supplement the chemistry set I received one birthday, which was ultimately used to etch the chromium off our kitchen sink with acids. The damage was so complete, so final, it was one of the few times I remember Dad looking in disbelief, too dumbfounded to give me the beating I deserved.

As is often the case, the true visionary behind Waddamana was not the government. In this case it was an intense and bespectacled metallurgist and inventor from NSW, James Hyndes Gillies, who came prospecting for locations to build an electricity generator in Tasmania in 1906. It was an earth-moving infrastructure undertaking of a style and scale more reminiscent of Isambard Kingdom Brunel or John August Roebling and their awe-inspiring bridges and tunnels. Seems that three propitious names are a prerequisite for being a prominent civil engineering entrepreneur. Gillies had invented and patented processing techniques for calcium carbide and zinc, both requiring lots of electricity, which was his driving motivation. The project was massive enough to require government approval and new legislation.

Regardless of his success at raising investment funds on trips back to London and acquiring the engineering materials from America, like my dad, a businessman he was not. Although it is debatable whether the political forces were deliberately set against him once the Tasmanian state government saw a cheap fire-sale opportunity for his patently brilliant plan. So, also like my dad, he ended up fighting the system, which in his case was the possibly unethical Tasmanian government, and losing. The double-dealing Labour government of the time was busy compiling a secret viability report on the project, clearly coveting it, at the same time as they were refusing Gillies additional time. Their very public investigation, probably with a duplicitous agendum, would not have helped either. Perhaps not so much business sense, as Gillies very adequately raised funds, gathered a talented team, and made considerable early progress on such a complex endeavour in such an inhospitable part of the planet, but more likely his not understanding the self-serving structure of political society. My dad hit his head against the brick wall of the same society and the machinations of politicians all his life. Machinations documented in Sun Tzu's *Art of War*, perfected by Niccolo Machiavelli in *The Prince*, and weapon of choice of Labour governments worldwide. It was another 20 years before the Tasmanian government saw fit to reward a much-deserved pension lifeline for the then destitute Gillies.

One never wins against such a force. Inevitably, the Hydro-Electric Power and Metallurgical Co. Ltd. was listed on the London stock exchange, and a division of Gillies' holding company, Complex Ores, became fatally financially stressed. A record cold snap in 1912 did not help in this terrain and weather, which is difficult at the best of times. The government's newly created Hydro Electric Department acquired his works in 1914 and the finished power station, later known as Waddamana A, was operational from 1916. It is indicative of how far Gillies had come in his eight years that the government capital could complete his work in two years and be operational, earning revenues. In my marketing workshops I often say, 'Everyone is lining up to be second.' It is a lesson about those sitting patiently on the side, watching and waiting for good work to collapse under lack of resources, to vulture the scraps despite having invested little themselves.

In my mind, Gillies deserves to be remembered more than he is, as the 'Father of the Tasmanian Hydro'. The project was ingeniously simple. Tasmania's largest lake, called, wait for it, Great Lake, emptied into the Shannon River high above the tree line on the plateau called The Steppes. The Shannon trickled a long way, eventually into the River Derwent, which is a deep, wide fjord bisecting the capital city of Hobart where I would also end up. A short way away from the Shannon River, at a much lower elevation in a very steep valley, was the Ouse River. A shortcut canal cut across The Steppes could conveniently divert the waters of the Shannon to precipitously fall to the Ouse. Falling water has energy, potential energy, potential to drive something like a water wheel or dynamo.

But someone has to do the work to transform vision to reality. Along comes Alexander McAulay, professor of mathematics and physics at the University of Tasmania, with his ability to determine how much water flows out of Great Lake along the Shannon River, and how much would be needed to turn turbines in the valley of the Ouse River, which also flows to the River Derwent, whose water in turn flows out of the massive estuary to the Great Southern Ocean, destined to circle Antarctica, so all was okay. A small masonry dam at Great Lake to increase and control capacity was the gambit; then a canal to a penstock holding pond, a downhill pipeline, then a bunch of turbines feeding power lines on pylons to Hobart. Most of it across Crown Land, so no private property to purchase and compensate, along trails cleared by my

Uncle David Branch. A small village of workers near the power station pops up and is called Waddamana, 'running water'.

McAuley loved the area. Rugged, isolated, and conveniently near his friends at the Wihareja property just off the unsealed country road to the lakes. Amidst the works, alongside the Shannon River and canal, he built a rustic but comfortable country lodge called Kanna Leena. This became the engineering and academic headquarters for the project, with all manner of interested political, industrial, financial and engineering parties visiting all the time including the particularly savvy John Butters, who managed to stay as the project's chief engineer under each of the project's ownerships as they changed. McAuley's son, also a professor at the university, and his fiancée were also visitors. But with rugged wilderness and complex terraforming comes risk and danger.

Interest in the property waned after the tragedy of the drowning of Margaret Kathleen Hogarth, the bride of McAuley's son Alexander Leicester McAuley Jr, on their honeymoon on 9 January 1931, while bathing in the Shannon River at Kanna Leena. By 1950, Kanna Leena was deserted. You wouldn't read about such a misfortune except in an autobiography.

When my newlywed father needed a place to live with his wife and a baby boy, we moved there, a couple of miles from his single men's quarters on the other side of the Penstock Lagoon. I was the baby, and this was my first home. The house burned down some time ago, but before it did, I visited there several times. The first time was with my dad, who showed me the primitive living conditions – no electricity, of course, or internal plumbing – the nursery in the tiny room at the rear of the house where my cot was located, and the by then miniscule Shannon River to the side of the property. With the power stations decommissioned, and the massive pipelines removed, there was no longer a need for diverting water.

My dad was an apprenticed carpenter with the Hydro Electric Commission, living initially at a desolate temporary construction camp called Hill Top, adjacent to the high Penstock Lagoon, and part of the team that constructed the wooden stave pipelines down to Waddamana B station. The camp was for single men, no more than boys some of them. The remnants are still there if you wander off the road and search amongst the brush. You'll find some foundations and prominent concrete steps leading nowhere – appropriate for this area known as The Steppes. Married men usually had a house in Waddamana village, at the bottom of the valley. When I was conceived and he married my mother, in that order, just in time, he had to find married men's accommodation, none of which was available in the village. By then, though, Kanna Leena was abandoned, and it was offered to him. So, my first home was Professor Andrew McAuley's country chalet, which had also been the first headquarters of the nascent hydroelectric scheme and the site of a horrifying drowning, on the trail of the canal from Great Lake to the Waddamana penstock.

These were academic giants, these people before me, but there were giants in my early life too. One of my first jobs was as a junior technician with the Department of Surgery at the University of Tasmania, operating along with Professor Robert Mitchell pioneering kidney transplants, and where I learned that the same Alexander Leicester McAuley Jr, when a professor at the University of Tasmania, was instrumental in the creation of an optics facility to supply precision sights and lenses during World War II, pioneering revolutionary optical design science. And I knew of Grote Reber, a USA scientist but also a researcher at the University of Tasmania, pioneer of radio astronomy, because I recalled his acres of arrays of poles and wire antennae at Llanherne adjacent to the Hobart Airport, and because he retired to Bothwell, last stop before Waddamana, where he was a local, albeit eccentric, celebrity, having a beer with my brother at the Castle Hotel. Or Dr June Olley, food scientist at the University but previously with the Australian national science body CSIRO, and with whom I published an academic paper on measuring the quality of fish. Today the university, like many others in Australia, seems to me to focus on enrolment fees from overseas students and little else.

There have been and are many Vandemonians. Most famous was probably Errol Flynn, born in Hobart to a biology professor at the University of Tasmania, to become the most prominent swashbuckling Hollywood actor of his era. Then there was Christopher Koch, who wrote the novel *The Year of Living Dangerously*, with its film adaptation starring Mel Gibson and Sigourney Weaver; Nan Chauncy, writing about children adventures; and Richard Flanagan, the Man Booker prize-winning author. Who else? Joseph Lyons, the only Tasmanian Prime Minister of Australia. Mary Donaldson, now Queen of Denmark. Bob Brown, founder of the world's first Green Party. Cricketers Ricky Ponting and David Boon, world champion swimmer Ariarne Titmus. Media celebrity Charles Woolley. Martyn Bryant, Australia's worst mass murderer. Truganini was the last full-blooded Tasmanian Aborigine, and perhaps the only true Vandemonian in this list.

CHAPTER 17

UNCLE DAVID AND THE ARIELS

There were four girls and two male children in my dad's family. He and his brother David were opposites. I am nephew to Uncle David and Auntie Marjorie, cousin to Greg, Andrew and Peter. David Branch was tall, slim, handsome in an aristocratic way, had a sharp mind, a slightly potty sense of humour, and was a bit intimidating.

I have known Uncle David since I was nought and he was about 15. David was quite a bit younger than my dad, who had just turned 20 when I was born. There are bits and pieces, but I have no sure early memory of Uncle David until a decade or so later when he was living in Tarraleah, another remote hydroelectric village taking advantage of another steep valley and high water reserves in the Tasmanian central highlands.

Basically, I was in awe of Uncle David. We never saw him as much as we saw his sisters – Molly, Leila, Phyl and Doreen – who all lived in Hobart so were closer to us. We saw him perhaps only once a year, and probably around Christmas time. I was in awe because even as a kid, I could feel that my dad was also in awe of his brother David.

The reason for that is important for understanding my own relationship with Uncle David. My dad, Reg Branch, was the flamboyant one, the dynamic, adventurous, bold one in the family, fearful of nobody. Yet my dad's precarious relationship with Uncle David back then was so tangible that even as a boy I could feel it and sense that it was out of character for Dad. The reason, I think, is that my dad had inner demons that prevented him from reaching his potential. Like many people with inner demons, Dad started well but then faltered. From the start, by comparison, Uncle David was a super achiever. David would have been the first to disagree with this assessment, but that is indicative of his self-effacement, his humility. I have a cherished picture of them together as teenagers that I call 'Dad and Dave'.

David had long, stable jobs. Dad struggled to keep a job. David married a beautiful, sophisticated woman from a hugely respected family. Auntie Margarine we kids called her, not understanding why the adults laughed. Dad's wife was more colourful and less predictable. David's career prospered and expanded into important management positions while my dad's career imploded. David was a great mechanic, excelled at go-carts and car racing. David always had a better house, a more modern car, and all of these things impacted on my dad, and therefore on me. Maybe the awe could more accurately be described as awe-inspiring. I remember nothing that my uncle did to create this in my dad. We all looked up to Uncle David. He was a class act.

When I was about 12 years old, I was unexpectedly called into the headmaster's office at Glenorchy Primary School – Glenorchy State School it was still called then. Something like this never happened, to any student, so it was very scary. The school was out of David Copperfield. Desks, one to a child, with wrought iron sidings, a lift-up top with obligatory carved initials from years of students imprinted into the wood, and a groove across the top for pencils to rest, beside the obsolete hole for the ink well. The headmaster was part of the administration of this large school, not part of the teaching staff, although he filled in when someone was away sick. He was elderly. When I arrived in his office I was met by a policeman – tall, slim, good-looking in an aristocratic way – and having not seen Uncle David very often and for some time, was surprised to see he was now a policeman and was at school to see me. Awesome. I said something like, 'Hello, Uncle David,' at which the headmaster and policeman looked at each other, then at me and back to each other again, completely puzzled.

Of course, he was not Uncle David, but had come as a result of a complaint from the property owner of a vacant block that I always passed on the way home from school. A day or two earlier a friend and I had sneaked in to grab some apples off a tree at the back of the block. The owner's house was next door and before we could do anything a savage dog came for us and we just managed to get back over the front fence to the safety of the street. Apparently, I dropped my school diary with my name in it, and the policeman, who to my juvenile mind looked like Uncle David, was here to give me a warning. The absolute last thing they would have anticipated would be what I said. With no training at the police academy for such situations, he was put off guard, and instead of being threatening to a 12-year-old as was intended, his routine became comic. I am sure the policeman has told this same story at dinner parties over the years. My unexpected reaction totally disarmed him and the headmaster, so nothing came of the matter. Thinking about it now as an adult I should have complained about the petty owner setting a savage dog on a couple of young kids just for taking some apples. But then, didn't most of us come out here from England for doing just that?

Eventually Uncle David moved permanently from the country, where he had been all his life, to the capital city Hobart, to work for the first time with a private company instead of the government, and where his relationship with all of the family blossomed. As I grew up and interacted with him more as an adult, my feelings towards him changed from cautious to appreciation. He turned out to be approachable, helpful, generous and fun. Slowly, there were more visits, and the shared love of music and dinner parties and family gatherings evolved. Two or three times I organised family get-togethers and he always came to them, joining in with gusto. I also went to his many house parties. After the death of my nan and then my dad, Uncle David became my surrogate, the substitute for the vacuum they left.

My Auntie Margarine was kind but aloof and I have few memories of her, or indeed the two of them together. I remember that she was dedicated to her three sons, particularly her middle boy, Andrew, who caught AIDS in the early years when it was still a death sentence, as it was for him. David and Marge were divorced. Auntie Marge could not survive Andrew's death, and she faded away with what I feel might have been a broken heart.

After Auntie Marge died, David married beautiful Ruth, who loved her music and social events as much as anyone. She was a party girl, and it was about then that I started to visit them in West Moonah more often, but still not frequently, partly because I was working most of the time overseas. Auntie Marge loved her music and partying too, but for her it was more within the bounds of her household, whereas Ruth wanted to go out and so did Uncle David.

My involvement intensified again when Ruth died and David hooked up with the exciting, sophisticated, accomplished Glenny Clarke. Hook-up is the right word, because in many ways Uncle David became more relaxed and more fun-loving than I had ever known him to be, and I could see that the two of them had closer interests than he had had in previous relationships. David endured with both Margorie and Ruth. Auntie Marge had died early through her despair and Ruth from a terminal illness, which would have affected Uncle David's perspective on life. Although Ruth liked country music and dancing, this became line dancing, which was not Uncle David's thing, and her social events moved to lawn bowls, also not a David Branch thing. I never saw Uncle David dance so much with his partner as he did with Glenny, nor get more involved in his beloved country music. David was a skilful dancer. I have favourite pictures of them dancing, not just at the country club but in the kitchen of their house at West Moonah before heading to the club. Like teenagers.

It must be a family trait because one of my favourite images of Mum and Dad is of them holding each other closely in the kitchen, dancing to a song that had just come on the radio, and totally ignorant of us kids around them. I even remember the song, 'Walk on By' by Leroy Van Dyke, who penned and sang the more famous 'Auctioneer' song. It is inconceivably difficult as an adult to reconcile images like that one with the others of violence, let alone what must have gone through my head as a child.

I think about these things. The complex contradictions of the people in my family. I wonder where it comes from. The progression of my dad and Uncle David in their lives as I lived mine along side them. Their highs and lows and frustrations and how they dealt or failed to deal with them. Other families I am close to are unlike this. In fact they seem kind of boring by comparison, stable and safe, but unexciting. Perhaps it is the mix of innate smarts with unsophisticated lineage throwing up shock waves.

Being not quite a generation older than me, David was part uncle, part brother, part friend. I'd like to think that he felt the same way about me, but we never talked about that. In time I was able to do things for him, instead of always visiting only when I had a problem for him to solve. He'd fix something mechanical; I'd fix something electronic. He'd play his latest favourite song; I'd introduce a different one. I'd house sit when they were off on their escapades. I enjoyed visiting them, more and more, and was delighted when I was specifically invited to his 75th and 80th birthday parties. They were the times I saw him with all of his friends, people I knew individually and had seen him with independently. Glen, Les, David Burgess, Llew, Mannix, his bike club mates, and so many others. He was in his element. David was a consistently loyal friend. During the last five to ten years of his life I felt I truly got close to him and knew and understood him. I loved him.

I remember the first time I went with David and Glenny to the country music club at Glenorchy, and watched them dancing, and seeing his mother 'Lil', my nan, in him, in his looks, his actions, his expressions. Nan was the closest one to me in the family, in my life, and David, without his knowing it, slipped into that place in my heart that night. Plus he was the last of my dad's generation, so he was precious to me for that reason too. He wouldn't have believed it, though; he was just always being himself and not understanding the impact he has on others. Had on others. His humility again.

Glenny is an accomplished singer and keyboardist, still very engaged with the music scene today, while Uncle David never sang or played a musical instrument. Nevertheless, he was surrounded by the top talent from our state and further afield. Several were his best friends. It was strange to me to see that mismatched relationship – musical talent with mechanical talent. His home always had musicians visiting, individuals who relished his company and hospitality. They gravitated towards him. It was rare for music instruments to be played at these sessions or for his musician friends to sing, they just drank and talked. I suspect that the musos enjoyed escaping from their ilk to a genuine friendship that was independent of that world, not dependent on their talent. Those types of relationships are always lopsided, but any relationship with David was balanced.

He wasn't as sentimental as me, so I could never express these thoughts to him; he'd laugh at me. And his laughs could bite, like the time when he laughed at my scalding my arm on a boiling car radiator. 'So how old are you?' he said. Yet the last thing he did for me was to modify one of his tools so that I could get a difficult nut off a wheel hub, never hesitating to help.

But actually, there was a sentimental side. He loved the music and jamming sessions after a night out, reminiscent of the family parties at Waddamana or Tarraleah. The music he loved was the sad music, the 'tear-jerkers', the sentimental country music. We'd reminisce at the parties. Tell each other over and over the same family tales and anecdotes, now and then a new memory emerging. More than once he said we were still making fantastic memories, right up to the end. Recently he described his memories from before he moved to the city of how he more or less cleared all the bush for the pylons to hold up the Hydro's transmission lines across the state with his bulldozer.

The one thing that I have heard so many people say about him, different from anything I have mentioned here, was that he was a gentleman. I always thought of him with respect, admiration, but it was interesting to hear not one, but everyone, say they saw him as a gentleman. That has left me with the same sense of awe at this special human being as I had when I was a kid.

David was a super mechanic and loved restoring old bikes. That was his passion above all, and he essentially lived in his garage working on them. He had a collection of a dozen or so of them at any one time: Ariels, England's Royal Enfields, BSAs, Matchless, JAWA, Triumph, Francis Barnett, any old bike. They were all British – never a USA Indian. He was an avid go-cart racer – the state's top champion – and an interstate runner-up champion, also dabbling in stock car racing. More than a mechanic he was a homegrown mechanical engineer, inventor, tinkerer, all in one, creating fabulously innovative solutions to mechanical problems, like putting the seemingly impossible traditional six-cylinder Holden engine in the tight east–west configuration of Morris 1100s.

Uncle David belonged to the Ariel Motorcycle Club, touring regularly around the country. In his 80s he was still riding his bike regularly. His coffin was adorned with Ariel mementoes at his funeral.

CHAPTER 18

NAN

Most of the skilled workers in Waddamana – the Aboriginal word means 'running water' – were refugees and immigrants from post-World War II Europe. Colourful people with pronounceable but unspellable Polish and Baltic names. I suspect that living and working in the village was a sort of halfway house for many of them until something better came along, so while there were the stalwart families, others came and went frequently. They were hard working, hard drinking, from a hard life, bringing something ethnic to the village, something the homegrown locals had never experienced, and would never otherwise have experienced.

The nearest village of any note was Bothwell, on its own little plateau halfway up to The Steppes at the top of the mountains, halfway from the coastal seashores. Unlike any of the otherwise indistinguishable midland towns across Tasmania, Bothwell was distinctive, memorable, tidy, sanctimonious, verdant, settled by Scots and therefore with its own golf course, the oldest in Australia, and home to sheep-herding landed gentry. The old money kind. Haggard old millionaires running around in ragged dungarees and clapped-out cars. Bothwell had the police station, Australia's oldest general store – which was to play a cruel part in my story as I departed from it one day to drive down the Little Den Hill and the nursing clinic. It was on the bus route to Hobart. Instead of a central aisle with a door at the front, the bus had rows of bench seats right across like the bench seats in older cars, with a dozen or so doors down each side to the outside, access for each seat, like cells of a prison. My great grandmother Fanny Branch had her family there, including my grandfather Russel Charles Branch, born in 1898, and whom we called Pa.

It proved impossible to extract much information about Fanny from my grandmother Thelma Lillian Harrex, who was born across the way at Osterley, in the Victoria Valley near the Ouse River where it finally emerges from the highlands. If there could be a village smaller than Lonnavale, Osterley would be it. I have a portrait photograph of Fanny and have viewed her grave site. She died a few months before I was born. I knew that she was beautiful and worked as a seamstress and dressmaker. But that was all I knew, and it was only later that I learned she had three children. In addition to Pa, she had my great uncle Cyril, whom I had met and was wary of, and great aunt Vera, whom I met only on her deathbed, afterwards joining the entourage of her three-hour funeral procession from Hobart to Osterly. Fanny had these three children out of wedlock to three different fathers. My prim and proper Nan would be too ashamed to discuss any of this skeleton-in-the-closet stuff with any of us as we were growing up. Once we learned this background, though, rather than shame, we all felt admiration for Fanny Branch and her breaking the staid Victorian mould. What's the point of being beautiful if you cannot have some fun?

Pa Branch's best friend was a mechanical virtuoso named Viktor Klicki. Viktor was married to Daisy, who was Nan's best friend. Both Viktor and Daisy were eccentric. Viktor got the coveted mechanical maintenance position at Waddamana Hydroelectric Power Station, and Pa, who was a skilled footballer and a skilled farrier and a hard-working labourer, was hired to go with him through the efforts of Viktor. So, they all went to Waddamana, *Grapes of Wrath*-style, cars breaking down and makeshift but insightful roadside repairs. Carrying bustling loads up the steep bends at Hermitage always a challenge, sometimes reversing to use the lowest ratio gear. A sight reminiscent of the horse-drawn carriages a generation earlier taking materials into the Waddamana construction site while it was being built and before they had any roads, hitched backwards on the wooden trainline rails to hold the load from running away from them when going downhill.

The two families – Nan and Pa's, and Viktor and Daisy's – lived next to each other in Waddamana, enduring through the pre-war years as Nan had her six children. My family of just the three of us at that time had been living at Kanna Leena out of desperation, but eventually a house became available in Waddamana village for us, a real house, smaller than Kanna Leena but with indoor plumbing, running hot water, a functional water closet, electricity, electric stove and oven, and close to Dad's parents at the other end of the sealed road at the back of the village. I was about two years old and it is where my first memories were laid down.

Even 30 or 40 years after the power station was turned on and the roads established, Waddamana was still isolated, so the village was as self-contained as possible. It had a general store, a playground park and swimming pool, a community centre acting also as a movie theatre and dance hall, a guesthouse, tennis court and cricket pitch, mechanical workshop, musicians and even its own Tacoma Narrows-like swinging bridge, which I loved to cross and jump up and down on in the middle to scare and thrill me at the same time. Years later I would often cross another bridge over Eggemoggin Reach, one-time home to Richard Buckminster Fuller and Seymour Papert, to holiday and sail at Deer Isle in Maine. It is the sister bridge to the infamous original Tacoma Narrows one. Galloping Gertie it was called; easy to see why if you've seen the terrifying videos of it swaying like a hula dancer in a storm, unbelievably with no casualties. The Maine one still operates, and although modified and reinforced several times, has to be shut down when the winds blow, which is all the time. Even on a calm day, crossing it feels precarious. Papert's Logo computer language was the inspiration behind my Tasmanian

Turtle educational robots. Buckminster Fuller invented the geodesic domes that dominate playgrounds.

Part of Waddamana being self-contained was having its own dairy. Fat, contented, lawn-munching dairy cows roamed the streets of Waddamana, held in by cattle grids at the road entrance and exit, built by Pa. They were attended to by a dedicated dairyman, Mr Mead, who lived on the corner house near my grandparents. Mr Mead would milk the cows before dawn each morning and visit every house to fill any pitcher, a pint, a gallon, that perched on front fence posts, as much fresh, warm milk as you wanted. Every morning Nan would collect her milk, scald the rich liquid on the stove, scoop the clotted cream from the top to use in deserts, and keep the freshly pasteurised milk for baking mostly. Nan, slightly plump, wrapped in an apron and sometimes a tea towel around her hair, was a wonderful cook. Had a reputation for it – published recipes – and later when my Pa had retired, she managed the busy Waddamana and then the Tarraleah guesthouses, catering to boarders, travellers and visiting VIPs. Manager, cook and cleaner. Proverbial 'chief cook and bottlewasher'. My favourite dessert, apart from vanilla Haagen-Dazs ice cream, is still preserved pears served in their syrup with mountains of clotted cream on top, which was a staple treat served by Nan.

Some cows had horns, some did not, and some were polled. They were a mixed herd of all breeds and colours. A cow with horns is frightening for a two or three-year-old. A cow without horns is frightening for a two or three-year-old! When I asked my nan what the horned ones were all about, she replied that they were to monitor badly behaved children. As I was a badly behaved child, the cows took on a heightened significance and I was terrified of them and always scampered out of their way, petrified if I came across them on the streets.

Parallel to the back road with my grandparents at one end, and our house at the other, was a scraggly foot track behind the houses, perhaps a dozen houses, so like a long back alley. Each house had a rear gate for access to the steep, wooded area behind, which rose suddenly up to the penstock and formed the sharp edge to the valley. Taking the track along the back of all the houses was a quick escape route for me to visit Nan, my favourite person in the world, without being seen by any neighbours or being caught by Mum.

I was the first grandchild, a little boy, and Nan nicknamed me Boysie. I loved that name. That is, until I was a teenager. Then it took ages for me to get everyone to stop calling me Boysie, such a childish name for a teenager. My Uncle David had friends with nicknames like Cork and Splinter, which I thought were more deservedly mine with a family name like Branch, until I discovered you cannot give yourself a nickname, it is bestowed upon you as a sign of merit.

As I wandered along the back track this day, probably thinking of canned pears and clotted cream, I noticed a group of cows ahead. They were ignoring me until I started crying, then they clearly became very intrigued by me. Somehow, they knew I was a naughty boy. I slowed down and unsure what to do, lost the opportunity to return because they approached me, surrounded me, deliberately, I believed, and I was petrified, in the sense that I turned to stone. I may have peed myself. In my little mind they were after me and I was rigid with fear. As I increased my sobbing, then bawling for real, the cows knew for sure I was guilty and came right up to me. Loomed over me. I was going to be speared by horns and eaten.

'I'll be a good boy. I'll be a good boy!' was what I screamed out to them between my tears. At the top of my voice I lied, 'I am a good boy!'

Nan must have heard me because she came to the rescue.

'I'll be a good boy. I'll be a good boy!' It wasn't the only promise I never managed to keep.

In those days in Waddamana, kids fended for themselves. It snowed in but it was very hot in this alpine valley in the middle of summer. The swimming pool was fed directly from water from the tail race of the power station, and as it had fallen a thousand feet and churned through hot turbines, it was warm water and we thrived in it. On later frequent visits to my grandparents' home, when we were slightly older, we would hang out at the swimming pool and playground with its swings, seesaw and roundabout. The vertical-board changing sheds were not entirely private and we tried sneaking peeks. None of us were monitored by adults, and none of us could swim very well, but we splashed in the shallow end of the pool watching more experienced kids diving off the slippery wooden springboard. This particular day my sister Sally, who was younger than me, started hopping up and down in the pool. She could not swim at all. The pool floor was not level and each time she hopped she took a further step down the underwater slope towards the deeper end, until her head was under water. She couldn't stop because of the slope and the momentum. Before anything went wrong, an older boy dived in and saved her from definite drowning. He pulled her out gushing and gurgling on the grass alongside the concrete edges of the pool. She didn't require CPR, which I am confident none of us would have been able to administer anyway. Years later I approached him, Peter Mead, the son of the dairyman, and relayed the story, which he had no memory of.

My parents never learned of this near tragedy. But there were other tragedies.

Tasmania had its share of flamboyant characters, not least of whom was Hollywood swashbuckling superstar Errol Flynn, born in Hobart, but there was also lesser-known Franklin Fausch. Just as swashbuckling and good looking. Instead of Hobart to America like Flynn, who sailed to Hollywood in 1933, Fausch went America to Hobart, arriving as a tall, entertaining, athletic wrestler in 1938. Fausch sounded a little too German, so after some personal rebranding, he took on the more likeable French name Francoise Fouché, and that is how he is remembered.

Hobartians loved him, women were intoxicated by him, men

admired him. He was exciting, exotic, worldly, patronising his health gym and frequenting his colourful night club called the Stage Door. Like many locals, my dad's oldest sister, Leila, fell in love with him. There were dance halls and night clubs in Hobart and a way of life that has disappeared, but I experienced the tail end of it. At 15, when I learned that the girl next door to me, Bronwyn, the object of my first crush, was taking ballroom dancing lessons, I joined up just so that I could hold her. I had no interest in ballroom dancing, only in Bronwyn. Our teacher was Alan Kutz. I have failed to find the correct spelling of his name, but everyone of the right age I mention this to knows the name. Alan was an international ballroom dancing champion and we were privileged to be taught by such a talent. Being able to dance has held me in good stead ever since. Bronwyn later boasted that she married an Alan, but not me.

Back to Fouché. The Stage Door was reminiscent of America's speakeasies: suspicious, dark, boozy; mysterious, smoky goings-on. Fouché killed a man there and was lucky to be acquitted of manslaughter charges. Lucky to avoid jail. But it changed his life; he was never to become quite the successful entrepreneur he wanted to be. The episode was followed by more police charges, failed real estate ventures.

Fouché spurned my Aunt Leila's love, and she then returned to her parents' house in Waddamana, set a rifle to her brow and blew her brains out. Well, partially at least. A rifle has a long barrel and if you hold the tip on your brow, the trigger is too far away, so she tied a piece of string to the trigger and the door handle of the room she was in, a guestroom separate from the house, and kicked it shut. She was sitting on a stool and the kick moved the rifle off target. She survived the attempt. Local shopkeeper Frank Cashion raced her to the nursing station in Bothwell, blood and brains from her frontal lobes loose, but saving her life. Leila survived into very old age, limping and speaking with a drawl, and had a somewhat altered temperament, but outliving Fouché and Kutz, who both died in car accidents. Nan, in her virtuous way, protected us from this dreadful story, as she did with what she would have believed was the unpleasant truth about her husband's mother, Fanny Branch.

I was with Nan at the beginning and I was with her at the end. I held her hand as she died. The first place I lived after leaving home was Nan's, and we lived together, more or less, on and off, for the rest of her life. When I was married, Nan stayed with us for some of that time at Lonnavale.

My dad was a violent man, a drinker, and to be feared. We would get beatings, sometimes too much. One time I was beaten so cruelly by Dad that I left home, went to my nan's, and stayed until it was safe to go home again. Another time he beat me with the kettle cord, three-pin plug forward, so hard, in front of his good friend Bobby Willmott, who had just informed him I had taken his car for a joyride. The beating was so severe that for years afterwards Bobby apologised to me, saying he never realised Dad was so vicious and he would never have told him if he'd known.

Dad appeared twice in court for gun-related charges, bailing people up. We never knew about those. When sober he behaved as if butter would not melt in his mouth, the stories only coming to light with later newspaper searches. I still do not know if he was ashamed or proud.

Nan and I were very close. Nan shared with me intimacies that were unusual for her to talk about, her post-Victorian generation with its prudery still redolent. She described her married life, her relationship with my grandfather, her childhood and dreams. When she was 73 I was working as a federal government employee in Hobart, living at her flat in Bathurst Street. I recall her walking up the hills after shopping in the city, struggling and age weary. One day I came home at lunchtime and Nan had a small suitcase packed. She explained that she wanted me to accompany her to her doctor's appointment, because the doctor was ignoring her complaints that something was wrong. It is just old age, was the ignorant comment each time. She said she wanted to be admitted to hospital for tests, and had her things ready for admission as an inpatient. She was desperate and felt someone stronger with medical knowledge would make them listen to her and take her complaint serious.

Nan's doctor was the wife of one of my lecturers from my medical school days at the University of Tasmania. That lecturer unfairly and inaccurately accused me of cheating one year at medical school because I wrote a creative essay to an exam question. This and one other similar incident probably contributed to my leaving medical school after three years. These were the arrogant but small-minded people in Tasmania that were my life until I moved to the larger world, where talent went up and egos went down. At the doctor's rooms I stated that my nan was not a complainer, and that if she felt something was wrong, then there was something wrong. Still the doctor refused to admit her to hospital, so that was a futile exercise, but she did take a blood sample for analysis. Amazing that even that basic treatment had been ignored up until then. And we went back home, suitcase in tow.

A few days later Nan received an urgent call from her doctor, and we went back again. Nan had advanced leukemia, which was causing the lack of energy. This time Nan was admitted immediately to hospital, and the next day as I was visiting, she lapsed into a coma and died. I held her hand as she left me, as I did with my dad a few years later, feeling their last breaths escaping this world. In one of our last conversations she laughed and said to me that here she was in old age, dying from a children's disease, which is what leukemia was known as in those days.

Before she died, Nan and I made a pact. She was softly God-fearing, and was not sure if there was an afterlife or not. She said that I was to try hard to look for a sign after she died. That if there was an afterlife, she would do whatever was possible to give me a sign, to make contact, to let me know. That was how close we were. For a few years afterwards I would look for the sign, at times allowing myself to be almost trancelike to try to feel her presence.

There was never a sign. Two years later was when I tried again while flying to Darwin to collect my baby sister Carol.

Like many families, our family life gravitated to one side, in our case my dad's side, the Branches. But I had a wonderful other nan, Nan McKenzie, my mum's mum. We saw her less often, and every time we did she brought us ice creams in dixie cups, a treat more expensive than I could normally afford. In those days the tiniest throat complaint led to a tonsillectomy, and when I had mine, Nan McKenzie brought me a dixie cup ice cream, as was the tradition post-tonsillectomy, but I was so sore, so out to it still, that I could not eat it and she gave it to the boy in the adjacent bed. I hated that boy. I remember the operation. I was maybe four years old, and I see a flask with fumes, so some primitive anaesthetic probably.

On Nan McKenzie's 90th birthday I returned to Hobart from an overseas project and was picked up and taken to the party at the nursing home by my Aunt Veronica. Nan McKenzie was almost stone deaf. I heard that people who lose their hearing fare worse than people who lose their sight. I cannot imagine not being able to hear my favourite music. Nan McKenzie had become withdrawn, internalised. When I walked into the room, the birthday girl was sitting alone in a chair, ignored by everyone, possibly ignoring everyone, as everyone around was chatting and eating, getting to know each other and work out how they fitted into the family tree. Deafness makes one internal, catatonic in appearance. Aunty Vonnie dragged me over and she yelled loudly into Nan's ear, 'Do you know who this is?' Nan could just hear if you yelled at the top of your voice right at her ear.

Her eyes opened wide, and a huge bright smile spread across her face, and she said, 'Yes, it's Boysie.' I've loved the nickname again ever since, but I was so effective at removing it that few remember it and it has not caught on again. Probably for the best.

CHAPTER 19

SMALL WORLD

The world can be extraordinarily small.

One of my friends in New York was the vivacious and beautiful Janet Wolfe. I was introduced to her by my great friend Faye Kilstein, whom I had met earlier on the beach at Santa Monica, on 1 January 1987 on Rose Bowl day, in California. The Rose Bowl Game is a traditional American football game held in Pasadena, California, each year and regarded as the 'Granddaddy' of the sport. It is a bit like the Melbourne Cup horserace in Australia, the race that stops a nation. Everyone listens to the Rose Bowl. The unlikely way in which I met Faye appears in another chapter.

Janet Wolfe was a socialite in the classical sense of the word; she knew everyone, and those she did not know knew her. We knew each other, instantly liked each other, and flirted innocently incessantly. In New York's predominantly Jewish Upper West

Side I built her some apartment furniture and fixtures; cut her perfect silver hair for her; slept in her salacious bed, alone; dined, cooked meals and invented recipes for her; met her two exciting daughters Alisa and Deborah; met her equally socialite friends; and entertained her with my Australian irreverence. In return, Janet never failed to make life better, funnier, larger than life. Janet and Faye were great friends, and she did everything she could to keep Faye and me together. When we first met, she asked me my name. I've always been a little sensitive to my family name, Branch, which I thought of as an ordinary, boring English word.

'Allan Branch,' I said.

'No, not your stage name, your real name.'

Hard not to love someone like that.

In New York, I mentioned to Janet one year that my company was looking for investors. She immediately put me in contact with two potential investors. After calling them and explaining the connection, I met with the first one, who turned out to be Australian and as we talked, I tested the Six Degrees of Separation thing. It turned out that we did, indeed, know someone in common. We both knew John Honey, a prominent Tasmanian filmmaker and writer. John won multiple awards, including the 1980 Australian Film Festival Award for his feature movie *Manganinnie*, from a novel by Beth Roberts who grew up at Bothwell, my great grandparents' village in Tasmania. It is about a runaway white girl found by an elderly Tasmanian Aboriginal whose husband had been murdered by white settlers in early colonial times. It was a clash of cultures during a typically abhorrent time now called the Black Death, that we learned about, unabridged, at school. We met when John contacted me about an educational TV show that he was thinking of. His idea was to use robots. From New York I contacted him and asked if I should do business with the potential investor and he said definitely not! So that was that. The other one, Julian, also turned out to be Australian, also with a Tasmanian link. He was from Melbourne but spent his school holidays growing up in Tasmania. That meeting progressed well; guests were invited to a large dinner party around the longest dining table I have ever seen in a private NY apartment. Straight out of a monastery refectory. During the protracted negotiations Julian returned to Victoria for a visit back home and died in a skiing accident, so that was the end of that. Janet had no idea these people had any Australian, let alone Tasmanian, connections, but it shows the world is small.

Janet lived to 101, a year after my last visit with her in the same apartment in New York, her daughter across the way. She was active till the last minute, although she'd started showing signs of memory loss. She was very clever at letting you think you were the only person in her world. Janet had a full life, initially working in the entertainment and movie industries, on first name basis with the likes of director Rossellini, Luciano Pavarotti of Three Tenors fame, and actress Shelley Winters. Janet was almost a lookalike for Shelley. She featured in almost 30 hilarious stories in the famous *The New Yorker* magazine's 'Talk of the Town', all from personal interviews by journalist Susan Lardner, such were her sensational exploits in New York, the city that never sleeps, the city that she loved. Self-effacing, jokes on her were her favourites. Just before her 100th birthday she said that for someone 99 she was doing extremely well, did not look a day over 98. No one I know was less involved with themselves than Janet. Janet did something amazing. A lover of classical music, she married a composer, and in 1972 she created and dedicated much of her life to the New York Housing Authority Symphony Orchestra, a 'full orchestra for aspiring and talented musicians of colour' who would otherwise have no opportunity. She was still serving it on her 100th birthday. It is hard to believe the world continues with her not in it. It is hard to believe that someone who socialised with the likes of Eartha Kitt, Mario Cuomo, Edward Koch, included me in her collection.

New York and its people became my default city, and I spent more time there than anywhere else. Someone asked me which was my favourite city in the world, and I said, 'I'm more or less living in New York and could stay here until I die.' They said, 'Knowing New York, you probably will.' Yet in all my time in the city I never witnessed one act of violence; in fact, there was more in my home state of Tasmania.

Sometime in the early 2000s I was hired by a Swiss medical technology company as its CEO. Elchrom Scientific was in the city of Zug, on the banks of egg-timer–shaped Lake Zug, south of Zurich. They manufactured electrophoretic gels for the medical research and health industries and were not growing, not profitable.

Electrophoretic gels are laboratory and pathology diagnostic tools to determine what macromolecules such as proteins or polysaccharides or DNA are in a sample substance like blood.

At the time the arrangements were being made, I was at Zack's in NY and as I was talking about my need to find a place to stay, he said he and his wife Marydel had befriended some people in Cham, a small town also on the shores of the lake and nearby to Zug, when touring Europe on their honeymoon. He tried but failed to find their contact information, which was by then decades old. Their friends were wealthy businesspeople and the friendship had been sufficient that Zack was confident that even now they would provide or find accommodation for me if Zack requested it. With no contact information, though, we had to give up on that idea.

It did not matter, because the main shareholder of Elchrom had a spare apartment in Cham, which was made available to me. The apartment was a simple and convenient one-bedroom flat, under a larger house, with a spectacular green lawn and a backyard leading down to the sometimes green, sometimes grey, sometimes blue depths of busy Lake Zug. The property had its own boatshed and runabout in a James Bond-like hidden alcove off the side of the lake, disappearing under tall willow trees. There was a small private pier and a diving area for swimming. At night floodlighting shone from within the willows, creating a glittering fairyland grotto to rival Peter Pan's Wonderland.

At times another shareholder would take us in that runabout for cruises around the lake, stopping off for lunches at exclusive restaurants, which had floating jetties for patrons arriving on boats. It was interesting to lurch up to a pier and fill the tank up with petrol.

Yet another shareholder made available to me his tiny pied-à-terre in the heart of Zurich's Old Town, looking over the towers of Grossmunster church, St Peter's giant clocktower, largest in Europe, and across the square from Fraumunster with its beautiful Marc Chagall–designed windows. The Christianity density was only exceeded by the vast number of commercial banks and sticky chocolatiers. Church and chocolate and cheques.

Elchrom was a private company founded by a husband-and-wife team, skilled scientists from the Balkans, who had invented a gel that was orders of magnitude better than anything on the market: more sensitive, more accurate, higher resolution, faster results, less expensive to manufacture, easier to use. The factory was in Zug, which had become a kind of Silicon Alley, a technology hub, in part because of tax incentives, about ten minutes away from my flat in Cham, which is pronounced like 'ham'. Zug is pronounced like 'hugue', not like 'hug'. They employed a dozen or so people and while the husband ran the laboratory and factory, the wife focussed on the sales. She regularly circuited the European and American universities, hospitals, research centres and clinics that had diagnostics, forensics, pathology labs and research projects. It was a simple and typical technology startup of the sort founded in a garage like Apple Computers. It was impressive but not growing like it should.

The founders were brilliant and dedicated, hard-working and motivated. Underfunded, like many startups of this type, they had attracted private investment and investors, including the ones providing me with places to live and lake jaunts. The investors were polite but were frustrated, impatient and pushy. They brought me in to the company to move things along.

I discovered a classic problem. The founders, while great technologists, were not businesspeople, and had limited marketing and business development skills or interest. They were unable to compete against huge multinationals like Fisher Scientific, despite having better product, so they defaulted to selling the gels at a discount to get sales. One of my tenets is that discounting your product devalues it. A premier product deserves a premier price, and customers expect it. Increasing the price was the key to reversing the company's commercial fortunes, and that is what I did.

One day one of the shareholders visited me in the apartment. He indicated that the largest shareholder lived next door and had asked us if we were interested in afternoon tea. It was a surprise that we had been next door to the main owner without realising it. We had admired the building – even in wealthy Switzerland it stood out. We had learned that they owned a dominant electrical components manufacturing firm.

At afternoon tea, the impeccably composed elderly couple, home bedecked with all the evidence of understated opulence – works of art, antique furniture – served us tea and cake on fine bone china while we made small talk. Their intention was to learn more of who we were. Who had intruded into their private company? The Swiss are extraordinarily private, nationalistic, and dislike outsiders, so they wanted to know who we were … who we really were. The talk got around to my residing mostly in New York, despite being Australian.

'Oh! We know someone in New York.'

'Really? I stay with friends of mine, artistic types, one an animation artist.'

'Our friend is an architect.'

'I know an architect, too, and often stay with his family as well.'

'We met our friend when he was touring Europe on his honeymoon.'

'His name would not be Zack Rosenfield, would it?'

CHAPTER 20

MUM

Is it possible to be neutral when describing one's mother?

Our mum was an unusual and special person. Barbara Branch, nee McKenzie, was among the most courageous, most determined, most generous people I have known. People thought they knew her, but probably really didn't know her very well. That was because she was also the most elusive and, yes, duplicitous person I knew. After all this time, I believe that trait prevented her real nature, one of big-heartedness, from being recognised.

It has been a great joy to rationalise that despite coming from humble and simple beginnings, Mum was far from a simple person. In fact, the creation of this chapter has proven complex and challenging as the richness of her life and the variety of each of my different experiences and sometimes conflicting memories have surfaced. Putting into perspective various comments over the years from others. Even now it is hard to characterise or summarise Mum's life, which is perhaps as it should be.

There are no memories of her, no image of her, at Waddamana. She would have been a girl, around 17 or 18 years of age. She was 17 when she had me, and I'd love to have a picture of her in my mind from that time. The earliest memory I have of my mum is at South Arm, still a very young woman maybe 20, when Dad was working on the farm for his brother-in-law Cedric Calvert, and we lived in a converted army hut. I remember her making homemade toffee and it sticking to the tray, set like concrete, how she laughed and how I enjoyed it as she banged it with a hammer and broken shards of wonderful sweet candy scattered all over the place. Over the years I have often tried, unsuccessfully, to make toffee for myself and for kids, and it is directly because of this earliest of memories. Mum was learning to cook, to be a mother, to be a wife, to be a housewife, to be an adult.

I remember playing hide-and-seek in the same old hut. Being placed on the shelf of a kitchen cupboard by Dad to hide from the others; there were two additional children by then. It was the inventiveness of the hiding place I remember, and again when I have played hide-and-seek with kids, it is a trick I have tried to duplicate, the most successful time hiding on top of a wardrobe from our dog Kari, who used to play hide-and-seek for real, but that doesn't count, I think. There was the time she came home with a baby who I always thought was my sister Olannah, but might have been Sally. I cannot get the chronology right and there is no one to ask now.

One thing Mum never did was drive a car. I remember her trying desperately to learn to drive at South Arm, finally giving up a few tries later when outside the Calvert property she managed to miss the whole gateway, miss the driveway, and crash into one of the gate posts. Those posts were the huge corner posts maybe a metre in diameter, used to brace the long farm fences leading off them. It had to be replaced! Dad was furious because of the damage to the car. At one time I helped each of the females in our family to get their driving licences, so with that success I tried to teach Mum. Suffice to say, again, that she never had a driving licence nor drove a car. It was a classic example of the elevator not going to the top floor. Somehow there was something missing in her ability to handle a vehicle and steer it. A conceptual thing, a genetic thing, a topographical inability. It is a rare thing, but I've heard of it since in other people. Yet she managed to get around more than any of us. She had about a dozen default chauffeurs, so I guess at the end of the day she won. It also meant that her physical strength lasted throughout her life because of the phenomenal amount of walking she did.

If I close my eyes and go back to South Arm, I can see her coming home from the maternity ward with Olannah, or was it Sally? Another great trick of hers, a trick repeated six times in total, plus I think a stillbirth in there as well. I even remember moving to South Arm from Waddamana, squashed in the front with Toni, in a truck stacked high with our things, arriving late at night with a bunch of chooks in the back most of which died from the trauma of the long trip. What I feel now from a distance of 70 years from South Arm is the fun and adventure my parents had, two young newlyweds, still kids really, starting to make their independent way in the world. Inexperienced but confident.

In recent times I have seen old black and white photos of Mum and Dad during these earlier days, and it is exactly as I imagined it. My imagination is of two people totally in love. One photo shows Mum, a happy and brassy 19-year old, with Dad looking at her, both playing in the saltbushes under sand dunes, and the affection is palpable between them. They are both glowing. I suspect Olannah, or Sally, is about six months from entering the world, and I am positive Mum is not aware yet.

This encapsulates, from my perspective, the most important thing in Mum's life, which was her love for Dad. They were two intelligent, hard-working, complex, difficult people, dealing constantly with internal demons that they inherited and which they were totally ill equipped to handle. But boy did they try! What they initially achieved against that background was immense. They built a life, a family, a financial base, and a network of involved friends and colleagues that lasted for over a decade before the same forces took it all away. But I really remember the fun growth phase: buying houses, carpentry, timber, tools, the smell of fresh wood shavings on the floor. And I remember wonderful parties, music, dancing. I remember Mum teaching me to waltz to Hank Williams songs. I remember Mum and Dad impulsively grabbing each other to dance when a great song came on the radio. This life of hard sweat and hard partying was like a rapid salvo of fireworks, like skyrockets each flying higher than the last, shooting from success to success. But it was a balanced life of work and play and I remember being extremely happy. There is a great photo of them one evening relaxing, arm in arm, at a cabaret, in all likelihood a reward after a week of labouring together tearing down walls and building new rooms.

Although I have many memories of Goulburn Street and Regent Street, including Mum taking me by the hand for my first day at each school, my next clear images are mostly at Hurlstone Crescent in the northern suburbs of Hobart, an unsophisticated working-class area more suited to them than where we had been in classy but pretentious upper Sandy Bay and where I suspect they never fitted in. This was 1960 and I was ten and it is where it all came unravelled. A bit like the simple country boy called Elvis being overwhelmed by the huge success he and others created around him. What I see here are two young adults, still out of their depth when it came to the rules and processes of life. They were unable to deal with the wealth and the kudos that they had built for themselves. They wanted to rest on their laurels too soon, party too early.

The hard work continued for awhile. It was a pity that it started to drop off, because I was at an age where the achievement and skill of the work started to mean something to me. For instance, my pride when Mrs Lobban, who lived at the end of the street, said she looked one day and Dad had started on a new carport and the next day it was all done and how amazed she was and what a great job it was. Dad concreting and fencing, Mum painting and decorating. One of my memories is being proud of Mum's work to decorate the boys' room in tones of blue, with our first matching bed covers and drapes, brocaded quilts as heavy as canvas but on simple tick mattresses stuffed with flock. The successes of my parents, their toil and rewards, drained them and they became more interested in the good life. The industry of their lives left them. Dad and Mum did renovate the interior of the house, but it was a more modern house than we had ever been in and did not need new rooms or new walls. Paint and curtains were enough. We got the ubiquitous swing in the backyard, but that was that. Dad built the carport and fenced the front yard, and that, too, seemed to be enough. Two photos epitomise this: one of Mum leaning against a power pole outside our house, the other of Dad leaning against his FE Holden panel van. There are no photographs of them together. The partying also continued, increased then dominated, and it was intoxicating, the music and the dancing, even a superman could not have escaped it. At that age they must have felt they were invincible, immortal, would live forever,

A suitable summary of this time might be that no matter what they did, their overpowering passions, negative or positive, came to the fore. I now remember Mum being alone much of the time, as Dad seemed often to be away, and when he came home, he was angry. During these times, with me not quite a teenager, I promised her the world, trying perhaps to be the man for her I thought my dad was not. Already Mum was becoming reminiscent, and she would sit around the fireplace fascinating us all with stories of her childhood. What I recognise now is that Mum still was just a young adult, maybe around 30 years old, asking for love and acceptance from her children; and my dad was trying to deal with changes in his life that he could not fathom. I think they were burned out. But even though their inevitable solution was to separate and go on with independent lives, they were still bonded in some magical way that showed up later.

The next set of images are through this period of estrangement, with a split family, and a period where it was my dad who became again my hero, a status he holds to this day, despite his flawed and human limitations. As a single dad, at a time when it was unheard of, he struggled to the best of his abilities and knowledge to bring up a bunch of overly energetic kids. This does not detract from my love of my mum. Mum never interfered with my loyalty to Dad, as much as it would have hurt her, and it was never within her to denounce him to any of us.

Eventually, though, after these two people had seasoned out and their lives stabilised, and the destructive fires had burnt to a smoulder, they remained each other's best friends. I secretly always wanted them to get back together, a child's naiveté, and something that never happened, would have been impossible, at least in the physical sense. But somehow, they got together in the psychological sense. It was my mum who was with my dad as he passed away. I have a sad photo at the very end of this period, taken at Grove Road, with Dad in obvious decline, but he is kissing Mum. There is a great scene in the movie *Inventing the Abbotts* where the son asks his widowed mother why she doesn't find a new man, and she tells him that there is someone in her life. Mistaking the point, her son says, 'Great. Who is it?' and his mother tells him it is his father still.

In 1981 my sister Carol, Mum's youngest child, died. As if it is happening now, I remember clearly her wailing like a banshee in her grief at the shock of the news. It was like the Middle Eastern Bedouins and their howling funeral rites, almost unnatural, and it was as hard to listen to as it was to bear Carol's death.

The pain had a cathartic effect, though. It precipitated her final transformation from a spirited and restless woman to a mature and sensible adult. Most of my memories are therefore of the last 25 years, with my mum loveable and likeable, helping bring up Carol's children Miah and Samantha, living on and off with Mim and myself, finally establishing herself with gusto as the matriarch of our family. The extent to which she changed her life and dedicated it to her children and grandchildren, especially Miah and Sam, during those years cannot be over emphasised.

During this period, I told anyone who was close enough to hear that my mum was my hero. In a way I had finally been able to recapture my lost parents and unite them in my inner self. One of my favourite photos is of Miah, Sam, Mim and me walking up Mt Wellington with Mum, young, fit and happy. A family group. From when they were two and four years old until they were mid-teens and left us, we were the family for Carol's kids. Uncle and Mim and Nanny, which is what they called my mum. Kindergarten, primary school, high school, weekend sports, broken bones, sleepovers, teachers' meetings, exams, homework, holidays. Then the family things, meals, eating as a family, cleaning, teeth, sicknesses, shopping, clothes, bedrooms, beds, reading stories at night, pets, lawns, gardens, Christmas and Easter, birthdays, parties, gifts, gift giving, aunts and uncles. All of it, my surrogate family. Sunday school, which was a mistake. In late adolescence they discovered religion and left us, never to be involved again.

During that time, Mum also built strong family relationships with her own children, and their partners, each individual and private to each of them, as indeed she had done with her own siblings, our McKenzie aunts and uncles. She moved from having a network of friends to her family being her friends. Each of them has relied intensely on Mum, which aided the bonding process because she was always there. Instead of out chasing adventures, she was now the substance of our family. Mostly she had been the halfway house, the hospice, as we have each gone through our own transformations and growing up. We recognised that, and appreciated it, and gave back as much as we could in love and in kind, but were never able to repay her enough for her support of us. She was not wealthy, so the aid was in the form of accumulated small things: a meal, a place to sleep, an ear to listen, company when it was most needed.

As I came to finally understanding who my mum was and how she ticked, I found renewed respect and love for her, and much admiration, as I understood the unusual strength she had to find to get through her life. Always without complaint. This renewed love came without any impetus from my mum, just from her being herself. The effort was all within me, and it was late in coming, because Mum, Sally, Dianna and Olannah were already like sisters when they were together. They obviously got there quicker than I did. There was rarely a photo of the daughters taken where Mum is not part of the troupe. No pictures with her boys, though, so I was making up for that.

And like true friends, she would have none of them 'bagging' the others. This is one of three key personal traits that have come to define Mum for me, this ability to 'keep mum' for each of her kids and to really love and enjoy each of them for who they were. It manifested itself through total support for each of us in whatever we were doing. They are traits that put her quite literally into the realm of really compassionate human beings and unconditionally loving mums. And such was her skill that I am positive each child believes their relationship with Mum has been the main one, a special one, apart from the others.

The second trait is related: it was her complete tolerance of people, of all peoples, in some ways so pure that it was naïve. There was not a single bigoted, hypocritical bone in her body. That does not mean she did not have a short fuse when it came to injustice or rudeness, she certainly did, but she literally loved everyone and was the most egalitarian person I know. It could be quite embarrassing at times, the way she would strike up a conversation with complete strangers. It is only when you saw the all-too-common opposite in people that you realised how special and rare it was.

The third trait was her generosity. I know of no other individual with the true generosity she showed to almost everyone. Most of our family will give if it is needed, but I think we would all agree that the giving is limited to what is left after our own needs are met. That is not selfishness, just common sense. Mum, on the contrary, would give the last she had and go without herself, without question and without hesitation. We would never know. It often left her privately working out afterwards how to handle some crisis as a result, how to pay the power bill, but to not give would never occur to her. Few people understood the degree to which this unbridled generosity sometimes created havoc in her personal life. For me, it answers the single remaining puzzle that plagued many members of my family when they tried to comprehend the dynamics of my parents' life. In a way, material things had no importance to her. In fact, she may have not had any comprehension of the importance and value of money. It certainly caused significant problems with my dad when they were married. If drunk and broke she would sell a new washing machine that Dad had bought, for the value of a bottle of wine. She would be unaware of people taking advantage of her, short-changing her, friends borrowing money with no intention of repaying it.

When I asked her about these traits, though, without detailing them like I have here, of course, she simply said that it made her feel good inside to behave that way. As a direct consequence of these three qualities, she was a person without an enemy in the world. Doubters, for sure, the unconvinced, definitely, but enemies, none.

I was able to discover these characteristics of my mother a year before she died, before I knew that I would need to recognise them. I was fortunate to spend an extended period living with her in her flat in Glenorchy, in some ways closer as friends than we have ever been. We watched movies, sang songs, concocted dishes, reminisced, and talked about philosophical things sons and mothers never talk about, like love, and life and death. It was a joy. My favourite recent photo is from this visit, taken in her garden, smiling handsomely while holding her dog Beau. Mum told me of her meeting Dad as he cruised by on a motorbike, of my conception, where it happened – first time and whammo: pregnant! I for one am really glad for that.

In a sense that encapsulates their lives for me. Everything in their lives was for the first time and they had to learn the lessons on the run, eventually succeeding, before the collapse, making mistakes along the way.

Several years ago, Mum underwent an operation to remove a tumour from one of her lungs. Mum, who had endured hard physical work and hard physical abuse without complaint, cried in her sleep from the pain. The surgery was deemed successful, but in 2005, after eight years of recovery, she learned that a new tumour, or a relapse of the original one, had appeared. Inoperable and with just a few months to live, she embarked on her final round of family visits, farewell talks and other private matters. She contemplated her own death. She volunteered that she'd love to be a fly on the wall at her own funeral. So, I wrote this eulogy for Mum and with her permission read it to her; in effect, letting her hear what would be said about her at her own funeral. She cried and said she had no idea. My sister Sally relocated to be with her and to help her through the illness, like she'd done during the convalescence from surgery the first time. Olannah and Dianna went out of their way to assist and to share as much time as possible with her, and the rest of her family increased the frequency of their visits. Olannah, in particular, showed amazing courage and consideration to host Mum, ensuring her visit to South Australia would be memorable, not matter how short the time might be.

Olannah's daughter Tia, always compassionate and empathetic, prepared an early Christmas dinner, one of Mum's favourite events, in October 2005, in case she was not around for the real thing. However, Christmas came and went and cheekily Mum enjoyed her second Christmas dinner for that year. New Year came and went. Easter came and went. By midyear, in between tours facilitated by her sister Veronica to her favourite locations around the state or to some she had always wanted to visit, weekend trips to markets and regular shopping and coffee sessions, she advised all that she would be around for the next Christmas too.

Well, as it turned out that was almost true. Just before Christmas Mum went into hospital for what was thought initially to be a troublesome chest infection. By midweek, though, it was clear that her cancer problem had entered its final stages. Our family was advised that she would not leave hospital this time and we gathered around again. Mum confided in Mim that she was now 'sick of being sick', and within a few days she had passed away, peacefully, not painlessly as she had hoped, but with great dignity. In keeping with her wonderful strength of character and her understanding of the effects this would have on each of us, her high-spirited acceptance of the end eased our own pain.

Mum was almost 74. She asked for a ceremony to be held in All Saints Church in South Hobart, the same church she was baptised in and married in, and for her ashes to be dispersed over her beloved South Hobart where she was born and where she spent most of her life and where she started her family.

So, as Mum would always say, 'Bye for now.'

CHAPTER 21

WINDOW DRESSING

This is the part of all memoirs where a decision has to be made about how much to reveal about relationships. This part is the whole intent of some memoirs, deliberate disclosures designed to shock. And sell. But they need recognisable names. Some memoirs are deliberately vague, or false, or self-serving. Others try to be as honest as possible, or practical as possible. Honest as possible is more revealing than practical as possible. I have elected to be as practical as possible. Being practical is safe, but enough. A bit like Hitchcock when he says it is the anticipation, when an audience knows more than the character, that generates suspense.

The first girl I fell in love with was a petite English import wearing a poncho, making her even more exotic, and a word I had to learn at the time. Because I was a college student and she was working, it contributed to my quitting school to be 'more mature' in her eyes, or so I thought. We were maybe 17. She dropped me for dropping classes, and it took a long time to adjust to being rejected. It was over while it was still platonic. Several years later, long after I thought that the breaking-up emotions were over, I glimpsed her on the footpath as I drove through the city, the first time since breaking up, and I had to pull over to compose myself, such was the unexpected reaction to seeing her. Then, many years later I made contact, had a brief chat at her place of work, and it was obvious it had burned out.

The next one was Sandra, losing each other's virginity, she 17, me 19, a serious romance of several years until something life-changing happened that determined the way I matured as an adult. Changed my very nature.

There are enough drinkers in my family that I have one of each kind. Dianna becomes a raconteur. We have all been uncontrollably convulsed at her anecdotes. Our Aunty Phyl, never married and with no children of her own, loved all of her nieces and nephews but she worshipped Dianna, who she called 'Baby', which puts my 'Boysie' to shame. My oldest sister Sally liked to feed people: 'Would you like a steak?' She was definitely our Holly Golightly: 'Home is where I feel at home.' Toni became verbally cantankerous, antagonistic. Even trying to avoid confrontation became confrontational. 'So, what, you want to just walk away?' Olannah liked to lament over music. Olannah, the smartest one in our family, intellectually, knew her way around the system in a way I wish Dad had known, but she also knew her music like no other. Others became amorous, or merry, or quiet, or danced, or slept. Mum liked to party; Dad liked to fight. For many of them one drink was rarely enough.

Growing up, I seemed to always be aware that Dad was aggressive. A bar-room brawler. It was only long after he had passed away, after a discussion with his younger brother Uncle David, that I learned the full extent of it. Uncle David said that Reg was a good fighter, but the problem was that he knew he was a good fighter. Made him fearless, unhesitating.

I never drank alcohol, so was always sober and in control. But emulating my father, I never shied away from a fight either. And I was good at it, I believe. Being like my father was what I thought I wanted to be. There is pride in some things Dad did with his aggression, like preventing the partner of his sister from beating her. Other things were shameful, like he and Mum and their constant domestic violence. In my emulation of him I ignored the ignoble and adopted the daring. At high school I was taken by my maths teacher to the outpatients to have stitches put into my knuckles from bleeding cuts from a schoolyard fight and to get a tetanus shot. Peter, the local high school bully, for reasons that will never be understood took a disliking to me and picked on me. One day walking past him he stuck his leg out and tripped me up. I spun around and flattened him with a single punch, then walked on. But I had to be taken to the hospital emergency department by a teacher for stitches to my cut knuckles. Afterwards his parents tried to sue me and the school for their son's broken tooth, but I was a school prefect and the school supported me. Many knuckle scars are all still visible across both hands; too many to worry about outpatient services.

Then one day during an argument I slapped Sandra, and she left me. I'm so ashamed I almost cannot write the words. I cannot even remember why I did it. I tried unsuccessfully to earn her trust back, but I learned that once was one time too often. Heartbroken again, I took stock of myself. Was this what I really was, who I wanted to be? No. I changed to be the passive, non-fighting person I am today. Bravado and bluff are still there and put to good use, but almost never anything physical. My sister Sally once said to me that I can hurt more with words than with fists. I put this skill to use in business negotiations these days; my lawyer has made me promise not to let him play poker with me.

At the time when I was dusting my knuckles at high school I was dating a wonderful girl called Helen, who a few years later lived in a flat next to mine. We had coffee together many mornings as we reminisced and she disclosed she had health problems. A few years more and this wonderful lady died from complications from the disability, still in her 20s. Unlike American schools, we didn't have coming-out proms, instead we had an end-of-year picnic – a day-long excursion on a remote beach on an island reached by chartered passenger ferry. At Dennes Point on Bruny Island at the last school picnic of my high school years, I was with Helen and she promised me I could have a 'feel'. I was still a virgin, so this was something special. We were to go into the surf past our waists so that it would be unseen. Before this lucky event, though, I got into a fight with the school boxing champion, and before we could decide if he was still the champion, roughing it behind the sand dunes, a teacher found us and sent us to the boat for the rest of the day. It was the French teacher, I think, Mrs Alexander, who has no idea what she ruined. Helen and I had been going around as a foursome with my mate Andrew and his girlfriend Gay.

A few years later I bumped into Gay when I was working repairing jukeboxes and pinball machines for The Automatic Music Company. She was in an office over the road, and we fell in love. This was before I went to medical school and Gay subsidised those years of my education, earning the right to be the doctor's wife one day. We bought into the property that burned down at Lonnavale then the property next door, and were married for seven years; to my mind six years of bliss and a final year of change. I dropped out of medical school and out of our marriage.

The last year of my marriage was one of confusion for me, and probably more so for Gay, I presume. I knew that things were different. Our years together had been so much fun and a true partnership, so the changes now were patently obvious. I still liked and respected my wife, but it was different. We still had a lot of fun together, so it took some time to realise what had changed: that I had fallen out of love. I did not know that that was even possible. Something primitive inside me felt that if you fell in love, you stayed in love. I cannot imagine what I understood about divorce and failed marriages, including my own parents.

Perhaps I considered those relationships to have not been loving ones to begin with. What they were I could not say, perhaps lust, or dependence, or just one-sided, unrequited. My marriage had definitely been a loving one, but it still managed to defeat us.

We had bought another property in the city, so I left to live in the property in the city with its mortgage, and Gay stayed on the farm at Lonnavale, which was debt free. The most troubling times of that life were when Gay would come to visit me in the city and she would burst into tears in public in the middle of the street pleading for me to come back. It broke her heart, and it broke my heart to have done it to her. I recall saying that as hard as it was now, in the future it will prove to be the best decision for both of us, which turned out to be true, as Gay went on to have a wonderful son from a healthy, caring relationship. I continued to visit her maybe once a year, on my flights back to Tasmania from whatever overseas engagement I was on, and it was when she gave me all the wedding photos that I knew our past now meant nothing to her. It is because I have been scared of committing again to anyone, not ever wanting to hurt anyone so deeply again, that I have never married again.

I first met Mim when I was still at Lonnavale. She was a student and lived with her family on a large dairy farm up the road. Later her first job was as a junior technician in the University of Tasmania Faculty of Medicine, in Anatomy, so there was something in common to talk about because I had worked in the faculty in Surgery, later to become a medical student at the same school. Because of that history I had been a visible student, so my name was known there. After separating from Gay, I met Mim in the city and we became close. Close enough that we stayed together for 40 years. But that too petered out. That is the only way to describe it. It was one of those relationship deemed to last forever, but surviving everything except time. We travelled and worked together around the world. After the first decade Mim began breeding dogs, and that prevented her continuing to be my companion on the overseas projects that had been our career and lifeblood. The 30 years after that are filled with memories of flying hundreds of thousands of miles in planes and living and working in exotic places like Paris, Chennai, New York and London, alone. It means that the other side of the coin of my adventure was a life of loneliness. I'd return to Tasmania at the completion of yet another project, and continue what was a more and more strained relationship. I'd catch up on overdue dog kennel maintenance, participate somewhat in Mim's dog programs by accompanying her and her friends to dog shows, attend veterinary visits, interact with the dogs, but all the time be searching for my next engagement. Then I'd head off again overseas where the big money was.

On two occasions I saved her life and, surprisingly, the events angered me. One time I actually applied the Heimlich manoeuvre, which like in movies brought up a bolus she was choking on. The other time I returned from an overseas project to find her exhausted, deflated, sapped of energy, expressing a desire to just give up. She recounted walking the paddocks with her dogs feeling so defeated she wouldn't care if she just died. One look at her and I recognised a goitre. She had hypothyroidism and the lack of thyroxin was sapping her energy, swelling her thyroid gland as it tried to make more hormone, the lack of which was causing her mood changes. An immediate trip to hospital for her thyroidectomy reversed everything for her. Mim recounted how she refused to take her iodine tablets handed out to everyone in Tasmanian schools, a state with such iodine-poor soil that goitre was a health crisis. This was before the days of mandatory supplementary iodine in staple foods. It also epitomises Mim, her non-conformance, her resistance to authority and refusal to change her mind even when logic dictates otherwise. It is possible that resistance to taking her iodine pills caused her goitre. One of my strongest memories from this time was my anger at Mim, for preferring her dogs over me, to the point where I felt like saying to her, so now, get the dogs to save your life. But, of course, I did not say those words. Despite this continued routine of regular visits back home, resumption of our conjugal life each time, there was no hint that we could return to how it had been as an international couple for the first decade. Mim wanted me in her life, but I wanted to be her life, not less important than her dog breeding activities. The failure to make a go of it with Mim is the greatest disappointment of my life.

Between overseas projects I didn't always return to Tasmania. Sometimes I'd stay with friends I'd made around the world, often from having been a house guest. I was in North Hollywood with Gordon and Dabni, having just secured a position with Nolan Bushnell in Sunnyvale, the heart of Silicon Valley. Nolan had founded Atari years earlier and had a passion for robots. He created a series of robotics companies, Androbot, Axlon, Chuck E Cheese. It had taken some time to negotiate getting this engagement, so ahead of starting it, I decided to head back to Tasmania for a short visit. But first, some R&R. A day before my due departure for Tasmania I drove to Santa Monica Beach, took a cassette player and a book, and laid out on a beach towel to enjoy some sun and relaxation, listening to a composite tape I had made. Three young, beautiful women pulled up at the concrete barrier of the carpark off the highway and walked down to the beach. One was wearing a full-length leg plaster, and she stayed sitting on the parking lot wall while her two friends strolled along the beach. I only partially noticed it all, and was content reading my book with earplugs in my ears.

At some point I realised the girl in the plaster cast was saying something to me, and when I took my earphones out, she asked, 'Who is winning?'

'What do you mean?'

She was immediately aware of my foreign accent, and she repeated, 'Who is winning?'

Not knowing what she was talking about, I walked up to her. She said her name was Faye. The most important sports match in America was playing right then and she assumed I was listening to the game.

'Winning what?'

'The Rose Bowl!' Incredulously.

'What is the Rose Bowl?'

'What?' More incredulously.

'What?'

A bizarre look came across her face as it dawned on her I really was an alien. She explained about this most important of college football games, as if to an infant.

'I thought you would be listening to the game,' she said, 'with the radio plugged into your ears.'

'It's just a music tape.'

'What kind of music?'

'Country.'

'Oh.'

We stayed chatting until her friends came back and rescued her. She asked if I'd like to have dinner the next day at a restaurant in LA she recommended. Unfortunately, I was flying out, so I asked if I could take a raincheck, and scratched my telephone number on her plaster cast. On my return from Australia a couple weeks later, she called and we had that dinner.

Faye was my first Jewish girlfriend. She was born in Memphis, of all places, and had a love of country music, so we had that in common. Who would have thought there was a Jewish population in Memphis? We dated in Hollywood for a time, me driving down from San Francisco for weekends. Faye explained about her rigid vegetarianism, from her being appalled at the carnage and cruelty at an abattoir at a kibbutz in Israel when she was stayed there. Later when she moved to New York, I stayed with her from time to time, helping her move apartments, meeting her friends like Janet Wolfe, and being indoctrinated to all things Jewish and all things New Yorkish. Culture, history, the Holocaust, Yiddish words, my first Broadway plays, kosher food, classical music. Faye and I are still friends to this day, despite that I never committed to her clear desire to get married. She described me as always calm and that she always felt safe with me. She was the daughter of a Holocaust survivor and my deep awareness of World War II and the Holocaust comes from knowing Faye. She visited me at various places around the world where I worked, and found the D-Day landing sites of Normandy with its liberation of Europe from the Nazis particularly distressing.

One day with her, driving my red Mazda RX7 through the Upper West Side of Manhattan, a pedestrian was deliberately walking slowly across our path against the red light. In New York, no pedestrian cares about the colour of lights, and half of the constant honking of horns is from motorists trying to assert their right of way and give advanced warning to strollers. Many are unaware they have even stepped into a crossing. Once made aware, pedestrians scoot out of the way. Not this one this day. I crept the car forward and just clipped the heel of his shoe with the front wheel as he sauntered lazily across. Faye was aghast, could not believe it, and said that she had never in all the years seen one act of aggression in me. I thought, if only she knew. As I moved the vehicle forward, the pedestrian turned and thumped his hand on the roof of my car, doing no damage but injuring his hand; in the rear-view mirror I saw him clutching his fist with his other hand.

My education and true awareness of World War II continued. In France, working for large appliance manufacturer Moulinex, at their European research centre in Caen in Normandy, I was exposed to the history and actual location of the D-Day landings along the Normandy Coast. More than that, though, the history of William the Conqueror, of the Bayeux Tapestry, of the evolution of church architecture and construction technology from Romanesque to Gothic, of flying buttresses, to the absolutely mind-blowing Le Mont St Michelle and its deadly, racing high tides on the English Channel coast. Tides so fast a racing horse cannot outrun them. Authorities send in helicopters to warn anyone way out to return to the banks.

As an expat in a foreign land, with just enough high school French to create laughter from my mistakes, I was fortunate to become friends with another person also working there at Moulinex, a German named Sabine. Two lost foreigners in France. Sabine was from Solingen in the industrial north-west of Germany. She is another very close personal friend, one of my closest friends, who has been there for me ever since. I have stayed at her house often when I am in Germany, initially in Solingen at her flat and at her parents'. Solingen is the home of quality knives and scissors and is at the centre of a triangle of the important

German cities Dusseldorf, Bonne and Cologne. Later I stayed with her after she moved to Freising in Bavaria, one-time home of Archbishop Ratzinger who became Pope Benedict XVI in 2005, and now in Munich adjacent to the expansive and enigmatically named English Garden, where we promenade like all Germans do.

Then there was Amy at CMU; Maggie, a prominent New York artist; and Melissa in Perth; Christine in Paris; Emily, an artist from North Carolina; Theresa and Louise in California. Leslie, a compassionate teacher in New York, who every time we meet walks with me through China Town, sipping bubble tea then dining on pizza and ice cream on a park bench. Any one of these amazing souls would have made a comfortable, loving, life-long relationship, but it would have been incomplete. I'd make it happen, but I'd be partly lying to them and to myself. A small number of prior times I have been so intensely in love, so totally captivated emotionally, so enthralled, out of control in love, that anything less would have been unsatisfying. Having the next work project to move on to has always provided me with the excuse to avoid admitting my inability to commit. But I had committed. I had committed to Mim and I was with Mim at the same time, reluctant for two decades to accept that it, too, was another failed relationship. Such was my desire for it to work with her.

Another time it was freezing cold, and having no clothes on did not help, but at least it was dark and I could creep away unseen, once far enough away so I could put the clothes that I was carrying all bundled up back on. Like a French farce: hearing the boyfriend come home unexpectedly, grabbing all my things in the dark, clutching them together and jumping through the, luckily, ground-floor window, worried about leaving behind a sock or footprint on the ledge, like an Agatha Christie clue. It added a new dimension to the expression 'window dressing'.

That time, being in bed with the girlfriend as the boyfriend unexpectedly and unwittingly arrived home, was not as severe as another time, another place, another ex-girlfriend, when I did not hear the boyfriend arrive. I was still in bed with her and somewhat surprised to look up and see him looking at us from the doorway. 'Somewhat surprised' is perhaps not quite the right expression. It was enough to ensure that I never got caught with the others.

One of my most fun projects was a company in Adelaide, designing and manufacturing Formula 1000 racing cars. Their operations were in Bangkok, and it was there that I met Jantarin, her rare first name meaning 'to brighten people's lives', with an unpronounceable family name. Like all Thais she has a nickname, 'Noi', which means 'little'. So, a bit like the endearment 'little one'. Not unlike my childhood nickname 'Boysie' or Dianna's 'Baby'. Even her family name is fitting, meaning 'joyful'. I've discovered over the years and over the continents that people are the same everywhere. That I would experience as close a rapport with someone in America, Asia, Europe as I would in Australia amazes me, but shouldn't. The rules of compassion, loving, cleanliness, politeness, honesty, integrity, tolerance, respect for others are, from my experience, universal. Mothers, grandmothers, aunts, older sisters, teachers, neighbours all instil these fundamental rules into kids everywhere. The puzzle for me is where and how it gets lost in the lives of those who head into the stratosphere of politics and big business. It sickens me, the lack of integrity in those leaders. Who were their mothers, their role models? Are they proud of them?

Noi grew up in the north of Thailand, in a family of poor rice paddy farmers. Her mother is a simple but beautiful farm worker who survived on a tiny family plot, a dedicated Buddhist. Her father, who never married her mother, was a professor of biology at a prominent Thai university and studied at one time in America. I met him and knew him and when he died, I flew over for his five-day Buddhist funeral, which was attended by representatives of the Thai royal family and the navy and deans from his university. Despite this simple, rural, working-class childhood, Noi broke the family cycle, an indication of her almost overwhelmingly powerful strength of character and dedication. She studied nursing at the best teaching hospital in Bangkok. It was a naval-based teaching hospital and Noi became a lieutenant in the Thai Navy, a rank she still holds. Then, as a qualified and practising nurse, she changed direction to build and run her own insurance brokerage business, and with it a successful life.

She never married, has no children, enjoys full ownership of her car and Bangkok home and everything in it. Noi encompasses all I believe to be desirable in my vision of the 'right' person, in my quest for the right relationship. She is honourable to a fault but not self-righteously religious; so, virtuous. She supports her family financially, paying her mother's way, putting her niece and nephews through school and university, securing mortgages for her other relatives. Stunningly beautiful, I hold her honesty, endeavour, intelligence, compassion, achievement, openness and pure heart in such high esteem. She has a certain innocence yet is street smart and not naïve. All are as important as the instant chemistry that existed between us from the start. It is not because I am running out of time – I'm not, and I'd easily be content with a friend or partner, which comes easily for me, rather than be alone. But Noi is the only one likely to alter my views on marrying again. She says I meet her ideal of a partner too. Most significantly, I am more important to her than anything else. I am trying to be worthy of her.

CHAPTER 22

REG

The first thing I did when taking the helm of Denning Mobile Robotics Inc in Boston was to sort out the fiasco from a robot sale they had made to Professor Robin Murphy at the Colorado School of Mines in Golden Colorado.

Initially Denning had developed a security guard robot called Sentry. They were exotic and expensive, and publicity was everywhere. The idea was the Sentry robots would patrol banks and museums and shopping malls on their omnidirectional three-wheeled chassis, using motion, fire, and other sensors to detect danger or foul play. Not a bad idea. As a competitor I was jealous of the promotion and visibility they were getting. Denning raised $20 million to list on NASDAQ in 1982, with the symbol GARD. A significant IPO for the times, probably equivalent to $300 million today. Hans Moravec, my colleague who I'd stay with when he invited me to Carnegie Mellon University in Pittsburgh, was a prominent technologist involved somewhat in developing their R&D and a member of their product development team. The technology worked, but the robots were unsuccessful and although some initial sales were made, the robots were returned and the large numbers of inventory robots that were constructed in anticipation of large sales had to be shelved. Fortunately for Denning, a second market appeared. Universities wanted rudimentary robotic vehicles with an open-source chassis or platform to conduct their research on. These research robots were not required to have any application, just the basics: motors, power source, computing system, and a menagerie of sensors that a scientists could tie into their own research project. So, all the security guard software from the Sentries was pulled out and the shell sold to universities as the MRV3. A few were sold, but the idea came too late because Denning, with no revenues, had used up its IPO money and was about to become insolvent. Just before Denning collapsed, an MRV3 was put together from spare parts and sold to Robin.

As soon as I was installed in Denning, I received a call from Robin saying her robot was not working and could I fix it. As usual I said, 'I can do that.'

I jumped in my Mazda RX7 and drove across the country to Colorado and fixed her robot. Robin was impressed that the new CEO of Denning was also a hands-on technical person, and that I pulled up in my car, took out my trusty tool bag, multimeter, oscilloscope, soldering iron and all, and got down and dirty. Robin's husband, Kevin, was a stay-at-home father. With an Irish name like Murphy, it was no surprise he carried a shock of red hair. He had chemistry degrees and was a very competent hands-on person himself, earning additional family income through carpentry, building decks and garden sheds and such. They had a young daughter Kate, who was born with a genetic defect so that her blood metabolites were out of kilter. In every way she was normal – smart, intelligent, energetic, beautiful – but Kevin constantly monitored her blood chemistry and adjusted whatever needed adjusting. It was a good arrangement. Robin pursuing her aspiring career as an increasingly senior academic in robotics, Kevin used his biochemistry knowledge to help his daughter alongside his extra tradesperson work. I was a house guest that time and many times after.

It seems every time I visited, I rolled up my sleeves. Another time I visited them and arrived from the plane from Australia with one of our Fander research robots in my suitcase, our competing product to Denning's MRV3. I took it out, turned it on and demonstrated it for her lab. She was stunned that it worked. This had never happened. She said typically salespeople have to arrive the day before, and set it up and tweak it before committing precariously to a demo. This had been what Denning had done.

We became good friends, and years later Robin was a full professor at Texas A&M University in College Station, in the middle of nowhere, centred between Dallas and Houston and

Austin, focussed on robotics for disaster events like flooding, building collapses, terrorist strikes. She established the Disaster Robotics Facility at Texas A&M, and her robots and robotics expertise were being used everywhere; at the World Trade Center after 9/11, for instance. Robin and Kevin had a second child, a boy they named Allan.

When she was in her late teens, Kate succumbed to her illness and died.

I would visit and have extended stayovers often. Kevin was a keen guitarist and singer of country music, so we would share versions of our various old songs. He liked Blue Grass and Southern Swing, and American singers such as Bill Monroe and Bob Wills; I liked Australia's Tex Morton, Reg Lindsay and Slim Dusty. At times I helped him with his carpentry work, and he would reward me by helping me select the tastiest ice cream from the local ice-creamery with their scores of choices. We regularly went to movies, and one time he and Allan took me to a model airplane fair, where the spectacle was a re-enactment of the famous X1 flight by Chuck Yeager, the first pilot to break the sound barrier, artistically portrayed in the movie of Tom Wolfe's novel *The Right Stuff* by Sam Shepard. In Robin's lab I'd continue to get down and dirty with the robots, one time getting a malfunctioning robot to work just in time for the PBS camera crew filming a story of girls studying in technical areas, or STEM.

Kevin spent a lot of his time in the kitchen being the chief cook, loving the chemistry knowledge as it applied to cooking.

He was meticulous, measuring quantities and temperatures and times. He was didactic, teaching me about the Maillard reaction, for example. I'm slap happy, mostly happy, when I cook. One day when I was visiting, from in the kitchen where Kevin was concocting something Cajun, he yelled out, 'Allan!'

There were two loud replies of 'What?' from the two Allans in the house.

Kevin called me over and said, 'We have to stop this confusion. I have an idea. What is your father's name?'

'Reg.'

Stunned silence. Eventually, though, Kevin said, 'What?'

'Reg.'

'Why, was Bruce taken?'

'What?'

'No one calls their child Reg.' He continued, 'Okay, I'll call you Reg from now on.' It was all agreed on.

Sometime later Robin called out, 'Reg,' and there were two replies of 'What?'

It turned out that Kevin's nickname was 'Red.'

In the beginning, my Reg, Dad, tried to be the world's best father. He played with us, taught us, included us, was proud of us. At South Arm, he drove me around the circular front driveway on his motorcycle. I had toy carpentry sets although I was only around three years old. There are amazing pictures of him and Mum playing on the long, white stretch of beach, me and my brother hoisted proudly above his shoulders, laughing widely.

Dad cared for me at times of my asthma stress. One time I was so breathless that he stayed by my bed, easing my fall into slumber. In those days it would have been a couple hours drive to the city, but no late-night medical facilities. When I woke that next morning my asthma attack had gone, and I remember asking Dad how. He told me he stayed through the night and pumped the only medicine available at that time, Aspaxadrene, from a complex glass atomiser with a rubber pump bulb, not unlike early fly spray atomisers, into my tiny lungs each time I struggled to take a breath. The relief of recovery next morning is one of my most vivid memories. Dad suffered the same form of asthma, explaining his compassion and patience.

More father and son memories at the next place at 41 Goulburn Street in the city heart of Hobart. He taught us how to make a graph to play Battleships and Cruisers. I loved drawing and he took my copied sketches of van Gough's *Irises* and an exploded car engine schematic to be displayed at the local general store, the community centre of our suburb. He played tricks on me. He looked at me one day, pointedly said hello, then walked upstairs to the second-floor bedroom where we children had our beds. I remember thinking it was strange, but did not consider it more. I was playing with coloured pencils and paper at the table with Mum, who apparently was in on the joke. A short while later he came up to me again from the back door said hello and made sure I watched him walk back up the stairs. The third time it took a cue from Mum before I realised that each time he came up to me to say hello he had not come back down the stairs. He had a tall extension ladder at the side of the house and came out through the second-floor window, down the ladder, returning quietly in through the back door. I was supposed to be surprised at how he mysteriously reappeared without coming back down, but his trick fell flat that day.

He walked us to the nearby Fitzroy Gardens to play ball games. It wasn't all games; he involved us in helping him with his work, but in a way that was like a game. He built a swing in each backyard; he made stilts and taught us to balance and walk on them. He let me play with his carpentry tools, one day telling me to be careful with his chisel which he had just sharpened, and sure enough I have a scar to this day on my left index finger where I cut it to the bone when it slipped. He had injuries too. One day while unloading lengths of timber or metal rods from his work truck they rolled on his foot and he limped around in a plaster cast for ages, him on wooden crutches and us on wooden stilts.

All the while he was renovating the house. He and Mum wanted a particular colour in the kitchen, so they played with tints and mixtures till they had the paint they visualised. My memory is of a deep blue-violet shade. I had asthma attacks, a lot, and any time an attack came on he would drive me across Hobart's spectacular mile-long floating pontoon bridge to Dr John Cannon on the eastern shore in Bellerive, never impatient or complaining of my illness. He was equally asthmatic. Dr Cannon specialised in asthma and had a new clinic, distinctive for its circular design, at the main road intersection.

The renovation was effective enough to allow us to buy a much larger, more modern, more impressive house in prestigious Sandy Bay. He built us a treehouse in the walnut tree, built billy carts with clever brakes and steering, bought us our first tricycles and scooters and pedal cars. One Christmas I had to follow a length of string all around the house and yard to find my present, ending up eventually at my first two-wheeler bike.

But Mum and Dad were alcoholics. I've heard that they were teenage alcoholics. When young and active and hard-working like they had been it had little effect on their efficacy as parents, as renovators, as partners. The hard work and sweat countered the effects of alcohol, or maybe hid the effects. But it took its toll. Dad's unmarried sister Phyllis lived with us and moved with us to the new house. Her boyfriend, my Uncle Lionel Hart, the one who raced motorbikes around the incredibly dangerous and daring Globe of Death, was with us much of the time too. My parents' time became dedicated more to themselves, and to music, dancing, parties, drinking. Mum and Dad would fight. Lionel and Phyl would fight. If Uncle Cedric and Aunty Doreen were around, they would fight.

It was still fun for the kids, more of us by then, and we learned to entertain ourselves more and more, walking down to the nearby beaches, or racing billy carts down Regent Street, or heading off to the movie matinees. School was important to me, although I was just an average student, and I still liked to draw. When only about six or seven years old I drew a sketch of a baby in a pram.

The teacher asked me what the bump was above the pram top and I replied that it was the blanket covering the baby. No one else showed any sign of an actual baby in their pram, just simple one-dimensional side elevations, but she gave me a gold star. Another teacher, I think Mrs Crocker, who we called Mrs Crocodile, had trouble with a student who could not get the spelling of some three-letter word right. I recall her getting impatient and angry and yelling at the poor kid. Today I know he was probably dyslexic as a reader and she was misplaced as a teacher.

But deterioration is gradual and takes a while to show. This house, at 28 Regent Street, was duly restored, value increased, a rumpus room added, the external bathroom integrated with the main house, the ubiquitous chook shed, the productive vegetable garden, and a treehouse for us in the walnut tree, new fence, a front patio with garden bed siding, new concrete driveway, and lattice work separators for the vegetable garden.

We then moved again, north to Glenorchy to 7 Hurlstone Crescent. This was not a restoration, and years later I realised that the covert reason for this move was so that Dad could be nearer his drinking buddies. Friends like Bobby Willmott, who he had befriended through one of the local hotel darts clubs. Partying was now overtaking his motivation to improve himself. Both Mum and Dad were having affairs. The drinking was perpetual. In the case of Mum, she was a closet drinker, with the usual talent for hiding her grog in amazing places, and in Dad's case, blatant drinking. June and Bobby Willmott had four girls and as Dad was around at their place a lot, we kids also hung out there, like a second family. The Willmott girls were like sisters and their mother, June, and another neighbour, Suzanne Lobban, remained very close friends ever since.

It was close to Christmas, and we knew there would be presents secreted in the house waiting to be handed out on Christmas morning. When I would arrive home from school, I was 11 or 12, as often as not there would be no one home. Searching through every nook and cranny I eventually found a guitar hidden behind the clothes in my parents' wardrobe. Every day I would pull it out. I cannot visualise it now, but it was most likely a small student's guitar in a cardboard box. I'd strum and play and be Elvis or Hank Williams, not knowing anything about tuning, then carefully put it away before anyone came home, ensuring it looked as if no one had touched it. All the time, each time, worried that someone would come home before I could put it away carefully, or that I'd mark the guitar or packaging in some way that would give me away.

Waiting for Christmas was always difficult for us kids, and I was beside myself with anticipation. Christmas morning came and I was doing my best to behave normally, not giving off signals I knew what was coming. And I was totally caught off guard when my brother Toni was handed the guitar. Toni. Atonal Toni, who loved music as much as I did, who he died as I was writing this memoir, who was incapable of holding the simplest tune. I tried my best to look surprised and then to look pleased when my present of a collection of bound classic books was handed over to me. A dozen or so classics, leather bound, gold edging, beautiful paper. Of course, I was starting to be seen as a good academic student by then and my wise parents were supporting that. I still have the books to this day and cherish them. I have owned a dozen or so guitars and now play at times in bands and sing, but that first one holds absolutely no memories other than this story. It was my nan, though, who encouraged me further with music as she did in everything. She bought me my first serious guitar, a Fender Coronado, from the manager of Allans Music (from the man who earlier had allowed a bunch of poor kids in to see *Ben-Hur* cheaply). Then this amazing woman bought me mail order guitar instructions, so by the time I was 13 or 14 I could strum a few cowboy notes and sing about as well as Toni.

The very last Christmas gift I received from Dad is something I hold in a special place in my house and my heart. It was in my mid-teens. We were still being cared for by Dad, but he was very much in decline, wasting hard-earned money on alcohol, which had progressed from beer to spirits. He would have struggled to buy anything for us, but the younger ones were easier. Just about anything is a good present for a kid. But I was older and more difficult to buy for. I received a cheap, plastic black-and-white shoe-buffing brush shaped like a penguin. It is by far the weirdest gift I have ever received, and a huge disappointment at the time. Dad would surely have seen the look of disappointment on my face. It was not long afterwards, when I was a little bit more mature, that I recognised the dilemma Dad would have had, with almost no money, struggling to buy something, anything, for each of us. Today it is a reminder of the extreme effort that our father went through to care for us and raise us, despite the enormous personal conflict he was going through.

Inevitably, Mum and Dad split up then divorced. With six kids – two boys followed by four girls – initially they agreed that Dad would take us boys and Mum would take the girls, but that proved too difficult for Mum and all but baby Carol came to live with Dad. As a single father he reared five of his children at a time when men did not do that. That alone, if nothing else, is how we remember him and appreciate him.

Mum and Dad became best friends later. A classic case of can't live with them, can't live without them. They loved each other till the end. Dad's end came sooner by 20 years than Mum's. He was in his final house, a government-subsidised residence, brand-new, so he could continue his handy work, establish yards with hedges and driveway, carport, gardens, terraced lawns and, of course, a chook shed.

Auntie Phyl died a few months before Dad, who died just a few short years after Carol, and Carol died a shorter number of years after Nan. My life and its participants were changing.

CHAPTER 23

LITTLE DEN HILL

In 1997 I quit robotics forever.

If robotics was my Middle Ages, then what was to come was to be my Renaissance. But like the real Renaissance, not before my own Dark Ages preceded it.

After 18 years of conceiving, inventing, designing, prototyping, perfecting, manufacturing and selling smart mobile robots, it was a difficult decision to leave the industry that I had pioneered, in which I had made an international name for myself. I had demonstrated that my background from a working-class family at the tail end of the world meant nothing, that it was performance that counted in the Silicon Valley world.

So, I had learned what I wanted to be. I didn't want to be rich, or at least not super wealthy like Jobs or Gates, but a certain fame and glamour was appreciated. I wanted recognition for a job well done, maybe. But that wasn't really it either. As simple as it sounds, the real buzz, the internal warmth I was getting throughout this period, was the feel-good of solving problems. Complex technology problems, business problems, human problems. For they are not separate. Particularly problems that smart people had tried unsuccessfully to solve before me. It was not to show off, despite that the projects were sometimes quite visible like the time the state premier attended a North American trade mission I convened for him, but because they often had a human element to them. Completing someone else's struggling project not only contributed to my kudos but also saved face for the other person. Successful problem-solving added to my knowledge and experience, but also developed those around me. Scientists, inventors, entrepreneurs, technologists, all learned from watching others, and they were watching me. At Commodore, after fixing a circuit design problem for my engineers when they said they could not solve it, I turned to see them exchanging money. I asked if they had bet on whether I could fix it or not. They replied no, they bet on how long it would take me. The same when I found a complex trigonometry bug in a robot navigation system that had defeated my frustrated and exasperated top software engineers. This became a feature of what was soon to come, but what came next was a surprise.

I was not yet 50 years old in 1997, but I thought an early retirement was in order. When I was still on the farm at Lonnavale, my goal was to retire at 30. It's not clear where that notion came from now. I do remember that an insurance broker who conned me into thinking he was the designated agent I had to buy my superannuation from when working at the university told me he had travelled around the world three times by the time he reached 30. That was a pretty exciting goal to my adolescent mind. Perhaps that influenced me. Retirement was an unknown concept really, especially when in my early 20s, so it was really a goal of working only when needed, being self-sufficient on my little hobby farm most of the time, with the odd project to earn some pocket money. But the house fire then the marriage break-up changed my situation and retirement became a lost thought.

But by now I had spun around the globe dozens of times, and seen things unimaginable to me back then, and I had plenty of money in the bank, so retirement was again a surfacing thought. Actually, not surfacing as in gradually bubbling to the top like water coming to the boil, but my immediate next goal. A decade of various experiences, followed by two decades of international technology leadership was to my mind a pretty good life career. I felt it was time to write about it. Reading had always been a serious preoccupation. Inside the front dust jacket of many books in my library are notes to myself of where and when the book was purchased. Typically something like 'Newark Airport, en route to London'. Mim said she would put the epigraph 'En Route' on my headstone. I said she should do it now then, 'In case I don't die first.' Among my favourite reads were biographies, especially autobiographies. *More Please* by Barry Humphries was especially memorable because of similarities to my childhood, billy carts and a stern father. Errol Flynn, David Niven, Richard Feynman, Frank McCourt, Ernest Hemmingway, Norman Mailer, Jonas Salk, Norbert Weiner, Harry Truman, Winston Churchill, Marie Curie, James Watson.

I headed back to Tasmania to settle down and write books. I owned a small retail business in the sleepy country town of Bothwell, where my Branch family ancestors were from and where my brother and a sister lived. Bothwell is village settled by Scots named after a Scottish village which was the site of a famous skirmish called the Battle of Bothwell Bridge over the Clyde River, which is also the name of the river through Bothwell, Tasmania.

My brother's name is Toni, not Tony, not Anthony, but Toni, named after an Italian friend of my dad's, Antonio. 'Toni', in English, is the female spelling and my brother has had a Johnny Cash, 'Boy Named Sue' life experience, sometimes receiving mail addressed to him as Mrs Toni Branch. Toni states a different origin to his name. He believes he was named after the midwife because Mum and Dad had not settled on a name by the time he was born.

Sister Olannah and her husband, Rod Davies, were aware of an opportunity to buy a rundown general store and felt that they could revive it. They wanted me to fund it and they would manage it. So, I became the owner of the Bothwell General Store and Newsagency, Australia's oldest continually operating general store. An Edwardian building, on the corner of a wide intersection with a residence above the shop. Unimpressive from the outside, but stacked like old general stores used to be on the inside and frequented by all the locals. I think Olannah was my parents' version of the Greek name Helena. I have wondered about my parents' name choices for their children. Drunk and misspelling? I regret not finding out where the choice for my name comes from. By the time Dianna came they did not even bother with a middle name, although it came back with Carol Jane, correct spellings even. Olannah had been married a few times and at one time her married name was Kennedy. She worked in middle management in a Federal Government department where she had to check some documents before being sent on to the director to okay. One day the director came down to her office rampaging about someone audaciously signing the documents off as OK; that was his job. They happened to be my sister's initials when she checked the documents before sending them to him for final approval.

With fresh funds to support restocking the store and the hard work of Olannah and Rod, the business flourished. Then one day Rod and Olannah appeared at my office to drop off the company delivery van that I had bought them. It transpired that they had won a competition for the most improved newsagency in the nation, such was the visible result of their hard work to industry observers. They were off on a holiday on the Queensland sunshine coast, and had no plans of returning. This led to a series of hired managers to run a business that really by then I had no need of. It also contributed to what was to emerge as my Standard Structure system of corporate management. But that was to come later.

Inevitably, and in character, I decided it would be interesting to run the business myself for a period ahead of selling it. I could continue my writing from there. So, I moved into Bothwell and delivered newspapers for a living. On a part-time basis I moved into the apartment upstairs, other times driving back to Wattle Hill and returning at 5am each morning. When I was about 11 I'd had a job selling newspapers door to door along Grove Road in Glenorchy, to earn pocket money from the tips. There was a competition at the end of each year with a prize to the delivery person who could complete their sales the quickest on the selected night. I raced around my allotted territory, sold all my papers quickly by letting everyone know I was in a competition, drawing on their generosity and compassion, and got back faster than normal to discover that someone's father had bought all the papers from one boy, who collected his prize almost immediately without leaving the newsagency, hours before I returned all expectantly. There was a lesson in that too. I describe it in my marketing workshops as 'It is who is first, not who is best'. Not as callous as it sounds, with special provisions applied.

Dusty country towns in midland Tasmania are indistinguishable from each other: a wide main street, some pubs, some fish and chip shops, some petrol stations, some stately homes, some shacks. Always on a river and if lucky, someone had the foresight to add a park alongside its banks. Bothwell is different. Some interstate visitors to my shop said that it stands out in their memory from past holidays touring Tasmania. Fifty years earlier I was a baby in a cot in the back room of Professor McAuley's tragic country house, and now I was the hit of the nearby town of my grandparents and those before them. My name was known, partly because of my family, but mostly from the media exposure I'd received as a 'Silicon Valley guru'. I did not know it initially, but many of my customers were not regular customers; they came to see who I was, to chat, some to flirt. Invitations to the expansive, old sandstone mansions, hidden away on acres of land behind protective groves of ancient trees, were constant. My brother Toni was indignant.

'I've been here for 20 years and never got as much as a single invitation, and here you are at dinner parties and guest of honour of families I never see, in homes I've never been to.'

Every few days I'd drive out of Bothwell, around the winding, mountainous road, back home to Wattle Hill. This day, at 6pm in July, the middle of Tasmania's winter, I closed up the shop, drove rapidly down the treacherous bends of the main hill, called Little Den Hill, so as not to confuse it with the other and higher Den Hill near Campbell Town, but still not so little, and careened straight off the road at one of the sharp bends, flying over and clearing the fence on the side of the road, rolling my car head over tail 360 degrees in the air at the steep siding, spinning once like a lawnmower engine that won't start or like a Catherine Wheel that started and stalled, Luckily and life savingly landing upright, ploughing through trees and breaking their trunks, continuing through dense scrub, missing a small dam of water deep enough to swallow the car, and coming to a halt way down the steep bank about 100 metres from the road. The time on my stalled wristwatch, which I have kept, is 6:17pm.

I woke up in the car, possibly only a few minutes later. I had been knocked out and concussed but otherwise only cuts and bruises. When I came to, all was black and deathly quiet. I managed to get out of the car, I think the door was open, not knowing where I was or what had happened. In the dark I heard a car engine up the hill, so realised where the road was, and scrambled through the bush back to the roadside. Disoriented, I wandered aimlessly. Soon a car came by in the dark and I tried to flag it down. The car kept going, but pulled up a short distance down the road and reversed to collect me. They could not take me on to my destination at Wattle Hill, but they decided the best thing would be to take me back to Bothwell to my brother's. When I asked why they did not stop at first, they said I was covered in blood and they were afraid I might be dangerous, but they then realised something must have happened to me. When I arrived at Toni's and saw my face and hair, saturated in blood and covered in encrusted dirt, I understood why.

For reasons I do not recall, the idea of going to hospital never occurred to us. I had no serious injuries, broken bones or deep gashes, I just wanted to get home. Toni could not drive me all the way home, but one of his sons would be going partway in an hour's time. After cleaning myself up and sitting around almost as if nothing had happened, but definitely dazed, I called Mim to explain what had happened and why I was late. We arranged that she would meet me halfway where Jason could drop me off on his way to the city. By the time I reached Mim things had become vague. I explained to her that I could not remember exactly what happened. It is a common symptom that the last few minutes before a tragedy like this that results in brain trauma become lost, as if the transfer from short-term memory to long-term storage did not have time to happen before becoming unconscious. The only memory I had was the car not taking the corner properly and the G-forces pushing me into the seat.

The next day I arranged for the manager to open and run the store, and Mim and I drove to my brother's place and later to see my car wreck. I still could not remember anything else. At Toni's everyone said they thought I must have fallen asleep because there were no skid marks on the road to show I tried braking. People had seen the car and had started talking about it, calling Toni. I did not accept that. I never drive sleepy, and besides it was early evening and I was not tired, but I had no alternative explanation. We then warily drove back down the road to see the wreck, tramping through the flattened path left by my car as it careened through the bushes. Every panel of the vehicle was smashed except the reinforced passenger compartment. If it were not a BMW, I would have died in the crash. Already someone had been there and attempted to take the car stereo and battery out. For many years the car – with its stereo and battery – was still there. As we walked around the crumpled wreck, the reason for the loss of control now became obvious: a blown rear left tyre.

No one would have been able to recover from the loss of control. Thankfully the car did not catch fire – I'd have burned to death while unconscious – or end up in the small pond, where I surely would have drowned.

About a half-year later I was in line at a voting booth in Bothwell, and the couple in front of me started talking.

'You look well; you've recovered from the crash.'

'Yes. Thank you. It is taking some time, no serious physical problems, but I was concussed and that has slowed me down.'

'What happened?' they asked.

'I'm not sure really. I know my back tyre blew out and I lost control.'

'We were the ones who you passed on the road just before the accident.'

'Along the straight at the top of the Den?' I asked. That had been some distance and time from the crash site, and I remembered passing a car on the straight section at the plateau of the Little Den Hill, somewhere between the Pub With No Beer and the start of the decline.

'No, that was the first car. We were the second car you passed.'

'Second car?"

'You passed us in your BMW, travelling fast but at the legal speed, and we both commented to each other about how well the car sat on the road. We came around the next corner and you should have been there, but you weren't. We drove up and down a few times looking for you but could not see or hear anything, so assumed you had either raced ahead or turned off quickly somehow.'

Suddenly the whole incident came rushing back to my memory. I remembered everything. I remembered pulling the wheel as hard as I could to come around the bend as the car refused to turn sharply enough, but the car just edging more and more off the bend. One memory flash was what must have been about halfway through the pinwheel flip through the air, when the force was pushing me down onto the seat and I was trying to stay upright, but then thinking that staying down would be safer if the car landed on its roof and caved in. Such practical thoughts in the midst of a crisis. Maybe the G-forces of the spin blacked me out, like those astronauts testing the limits of their endurance. I also remember not panicking, thinking calmly that this is how I am going to die. The whole crash would have taken seconds. Afterwards I thought about the pilots who struggle to recover a crashing plane, their voice recordings calm as they do the designated manoeuvres to recover from the disaster. Not thinking or talking about death, no screaming or panic, just intense focus. The road noise I first heard when climbing out of the crumpled door of my car must have been them going back and forth along that short section of road looking for me. It might have appeared to them like how the DeLorean disappeared in *Back to the Future*. One minute it is there, the next it is not. It took me not to the future but back to the past, back to dreams of retirement gone.

Just as Carol's death had affected me mentally, this affliction dramatically and drastically physically altered my life, this time permanently. In time I recovered from the anguish of the death of my sister, but I have not recovered physically from the car accident. I felt not like a shattered vase glued back together with visible cracks, but more like a vase reconstructed from the shards, with pieces missing, part of the bowl missing so it no longer held water, half a handle so it cannot be held up, sides missing from the base so it does not stand straight. That was me after the car crash on Little Den Hill.

Not surprisingly, there was a psychological aspect to the car accident. It crept up on me. I lacked energy, incentive, drive. It took a year to recover from the mental trauma of concussion, I was bedridden for much of it to avoid the world, avoid reality, unable to cope or manage my finances, treated with antidepressants to accelerate my recovery, experiencing financial meltdown, to use Zack Rosenfield's phrase.

It was the end of my shop days – I let it go – and the end of my writing aspirations. Coming out of it, with expert help from my doctor, I needed to find work again and like my friend Zack, who could not go back to architecture, having recently abandoned what I had been doing for almost two decades, I could not go back to robotics. At this time I made contact with my friend Joe Engelberger, the father of robotics, who was now in California, having sold his Helpmate in Newton, Connecticut. He said, 'We all wondered what had happened to you!' I could not bear to tell him I had abandoned robotics.

In time there appeared more physical damage than initially realised. The day before the accident, I could run up stairs three, four, even five steps at a time. The day after and forever after I could not even run, could not spring or jump, could not balance if I closed my eyes or leaned over. There had been nerve damage in my lower spine. Climbing stairs from then on was a step at a time, tedious. Getting my life back was a step at a time.

During my recovery and convalescence, I eventually started regaining motivation and energy and remember going down to the end of the long drive by the country road to wait for Mim to return home from her work. Our dog Kari would be by my side and our pet sheep Abba, as in 'a baaa', would join us. Mim said it was fantastic to come home and see her man with his dog and a sheep waiting for her at the farm gate.

A year of letting my finances collapse meant I now needed to find a job. Composing a resume became a work of art. Having been founder and CEO of my own technology company for so long, not progressing in a traditional career path, meant I had no CV. Had never needed one. I was rejected for the first couple of positions I applied for in Tasmania. That was hard to accept. A few years later I helped one of the recruiters with their master's degree and a business plan for his recruitment firm. We met originally because he was recruiting for one of those Tasmanian clients, and he admitted then that he wanted to put me up for his client but there was something too intense about me. Now, like looking down on oneself in an astral dream, I can visualise back at myself and recognise that I was still affected by the concussion and medical treatment for depression. Functional again by then but with some residual symptoms nevertheless. Today I stop and take stock of my character now and then, concerned that I have been permanently altered in some way, my new balancing act.

Work and money were now critical. Now came my collision with the international corporate and finance and investment world, ultimately ending up in Wall Street.

With no luck in my home state, I moved to Sydney and stayed with my sister Sally in Auburn, Sydney's ethnic west, in an upstairs apartment so close to the railway tracks that it felt like the train was going through the bedroom at night. Of all things above an undertaker's, with a solemn chapel, piped dirges, a body preparation room and a stack of refrigerated mortuary storage spaces like the large slide-out shelves seen in the movies. I got some work helping Proctor and Gamble recover from a messy software upgrade. Then I got my first new real job: general manager of an accounting firm called True North Group, or TNG, in the city. True to stereotype, the undertaker's assistant, attractive and eerily Gothic, who worked in the office downstairs, one day offered to dress me up and apply make-up. It looked promising in a Munster's kind of way and I was tempted until she said she'd do it in a coffin, a fetish that appealed to her but did not really appeal to me.

One night before I moved there with her, Sally had come home on a bender, having been out partying all night in Paramatta, but could not get in because she had locked herself out of her second floor apartment. That means she was probably already on a roll before leaving to go out, forgetting her keys like that. She managed to clamber up to the roof of the carport in order to try one of the windows accessed from the driveway side of her flat, but everything was locked. Most likely she climbed on top of the body storage fridges which were outside of the main building but secure down the driveway. She settled down and fell asleep on the roof top and was let in next morning when the undertakers came to work. It is hard to imagine what they must have thought, who would have been the more surprised. She said that during the night she had an epiphany from God. She actually used that word. From that moment on, so real was her vision that she became a dedicated born again Christian. It never stopped her from the occasional bender and party.

True North Group was a prominent business, in the heart of Sydney's CBD, and provided a complete suite of financial and accounting packages for personal and corporate clients. Some services such as succession planning had never made a profit and were offered as part of a holistic package for their clients. TNG was the beginning of my collision with the international corporate world. I began discovering rules and tools and techniques about companies and their management that were at variance from accepted wisdom, just as I had discovered in the robotics world. It started in Sydney. TNG was run by a very accomplished accountant family, Ian and Mark Bruce. Quintessentially professional, extensively connected in this financial centre of Australia, A-list clients. For me it was a risk to take the position, scary, not a technology company, which is all I really knew, not even a manufacturer, but a financial services company, and a large mature one at that. Taking on management of someone else's business was not an issue, and the financial accounting technicalities were no problem as I'd restored an insolvent NASDAQ-listed company in the USA. It was my not being the CEO that scared me. I had always made the decisions and called the shots, taken responsibility, answered to no one.

Within a short time I had shown an aptitude for analysing the business and those loss-making divisions started to become profitable. I was promoted to NSW CEO, then National CEO then Global CEO in quick succession. When asked by the owners how I did it, I replied that I really did not know. And I didn't. All I did was duplicate what I had always done in my private company. It seemed to work. I was learning that building business at the top executive level was independent of the industry. Planning, strategy, vision, leadership were requirements; however, specific industry knowledge was all but irrelevant, and designated to the general managers. More than that, I realised for the first time that this high level trait was essential, not arbitrary. This came about because I recognised that the family of skilled accountants, who had been running the company, were intelligent, dedicated, but not executive management material. Later I was able to generalise that professional specialists like accountants, lawyers, consultants and scientists rarely made good executive managers, and is why founders were often replaced in their startups. The best entrepreneurs were not the inventors who said, 'How clever is that?' It was the person who said, 'What can I sell today?' I discovered what became my Standard Structure.

Without knowing it, I was transitioning from a person needing a job, needing to work to earn some money, to a person discovering hidden skills other than the technical ones that dominated my past. Skills that were always there, always being used, but invisible behind the universal interest in the glamour of the exotic robotics technology.

This was at the height of the infamous dotcom era, and one of the TNG clients, a property developer, architect and builder called PlanetBuild, wanted to join in all the digital online fun that was ravishing the world and attracting millions of effortless investment dollars. Their offshoot, eBuilt, was incorporated and I wrote their business plan to raise seed capital of a million or so dollars, flew to San Francisco with the owner Drew Innis, and did a mini road show. Within 30 days I had raised their money. It was as easy as had been reported. The keystone investor asked me what I was going to do next, and I said fly back and write the Round A business plan for $12 million. They said, 'Stay here and write it and we will participate in that.' I bumped into the amazing Stephanie Race while using the internet at a library, and she joined the road show as part of the company. The business plan was dated 7th July 2000. The dotcom bubble burst in March 2000. The debacle that the dotcom phenomenon was to become had just started to arrive, as inevitable as it can be seen now in retrospect, with college dropouts running around espousing that 'the rules of business no longer applied to them'. The next road show was much more difficult, but those dotcom companies with their growth over profit models all fell over like dominoes. I was watching and learning from the inside. It was about to be a rich hunting ground for me.

At TNG came the embodiment of my tenet about accountants not suitable for managing a business, adding to my emerging knowledge of corporate structure, while the dotcom project showed me I could write and sell successful business plans, even as the easy money was drying up. This is where my nascent thoughts on Standard Structure first matured into a full-fledged theory. All it entailed was what I had been doing for years, what came naturally from managing and growing my own company. I realised that, if the same management style worked equally well for an advanced technology enterprise as it did for a mature financial services business, then running a company was independent of the industry. And again this contradicted the accepted wisdom. Even now, 20 years later, when I hear a struggling company seeking a new CEO, espousing that they need an industry expert, I have to point out that it was an industry expert that got them into trouble. What they need now is a company expert.

So, my first companies out of robotics were a financial services company and an internet services company. More came: an Indian dotcom in Chennai, a medical products manufacturer in Wales, a feature animation company in Hollywood, a nanotech startup, a genetics researcher, a glass artist's workshop like Warhol's Factory, a mining company, a tourism precinct, a telco, a travel agency, a radio station, a racing car manufacturer in Bangkok, a banana plantation waste processing company in Egypt. At some point it all led to my basing myself between Germany with my friend Sabine and New York with friends Gordon and Dabni. In New York, between projects, I was working with my very best friend Zack Rosenfield to transition his business. Zack told me afterwards that saving his business saved his life, so suicidal had he become from the financial meltdown and hopeless struggle to resolve it. So, I have added that to one of the benefits I get from turning around companies: it saves investments, mortgages, jobs, reputations, and now and then, lives.

This definitive turnaround skill set evolved and put into practice across companies and industries, across the world, with all manner of problems, became the next phase of my life and career, a huge change triggered by missing that corner on the Little Den Hill. But robotics lingered just a bit in my soul.

While in Germany in 2004, I read an article by one of my heroes, Rodney Brooks, an Australian from Adelaide who was a robotics professor and director of the artificial intelligence activities at MIT. It was in the popular science magazine, called *MIT Technology Review*, in the German edition. Along with scientists like Hans Moravec, Rodney was one of the premier robotics pioneers and key in promoting that the evolution of intelligence necessarily depends on what he described as representational interaction with the environment, which necessitates motion: the need to solve problems involving senses and actuators. I believe that. There are no intelligent cabbages, for instance. That said, for a plant, coping with the world is dramatically more difficult than for an animal. An animal with the genes to allow it to run away from survival threats solves most problems. A plant has to do it while standing still. Plants, therefore, usually have much larger genomes than animals. A human has about 24,000 genes, an onion about 240,000. But few would consider the very competent coping mechanisms of a plant in its environment as intelligence. Smart, clever, elegant maybe, but not intelligent in the sense we usually think of it. All the plant solutions are hardwired through regulatory genes. No cogitating on a never-before-thought-of solution to a problem. The onion is not thinking through a problem.

Rod's article in *MIT Technology Review* speculated that in ten years' time robotics would come of age. In fact, shortly afterwards, Rod left MIT and created his own new robotics company called Rethink Robotics. I and others had said the same thing over and over again, but it never happened. Now divorced from the industry, I was tired of hearing it. It was the failure for this to happen that precipitated my decision to quit robotics. So again, being who I am, and despite not being in robotics any more, I needed to work out why, going back to basics as always. In the process, I discovered two things. One was the pricing constraints of robotic products and services; the other was an interesting, previously undocumented pattern within the industry. I contacted the editor of Rodney's article, then with their permission wrote and published a follow-up article for the magazine, in German, called 'The 15-Year Cycle of Robotics'. It has also been presented in conferences and guest lectures.

In my hometown of Hobart in Australia lived an expat American named, get this, John English. John was the most successful author in Australia of reference books for small to medium sized businesses for publishing house Allen and Unwin. John knew me and was intrigued with my work building an advanced technology company and, more recently, turning around businesses. After some chats over coffee, he introduced me to Patrick Gallagher, the owner and manager of Allen and Unwin in Sydney. An unbelievable contact.

He said to Patrick, 'You should see this guy, he has books in him.'

When I met Patrick I said I had several ideas for books. A coffee table picture book on robotics, an autobiography on Allan Branch and robotics, a reference cum textbook on robotics, a book on turning around companies, and I had an idea for an industrial espionage novel involving genetics. Patrick deflated me. He said they would have no interest in all that stuff about Allan Branch and robots, but the turnaround book and the novel sounded like good ideas.

Ironic that the one I am actually writing is the one about me.

CHAPTER 24

JAKE AND JUDE

One of the first records I bought, after 'In the Misty Moonlight', was an Elvis song called simply 'Judy'. It is a little-known song, easy rockabilly feel, short and sweet. I still like the song and play it on YouTube from time to time – easier than putting the original 7-inch vinyl 45prm disc into a mechanical machine. Not that I have that record anymore, all of my original collection being destroyed in the house fire at Lonnavale.

There seem to be a few Judys in my life. My first overseas trip, to Auckland, was to escape one; then there is Judy Goldberg in New York and mother of great friend Leslie who tours China Town with me; most recently there's new-found friend Judi, who helped me out of a difficult spot and dances with me at music clubs, spelled as in Judi Dench.

While working for Krucible, a junior explorer mining company in Townsville, in the far north of Queensland, I bought a two-bedroom unit. It looked for awhile that I would remain in Townsville, given that after three years of the most difficult engagement of my turnaround years, I had finally saved the company from failure, raised record investments, stabilised and started regrowing its share price from years of decline, and initiated a growth and expansion plan of acquisitions and asset building. This was during the Global Financial Crisis, when there was little to no investment, and at the height of the worst mining downturn in recent memory, when there was zero interest in mining companies, least of all explorers. Much of my method involved promoting the company on the international conference circuit, which kept me out of town for extended periods. The presentations at these industry conferences attracted the investments, secured our position as a prime rare earths resources company, and even resulted in published research papers.

Being away, though, meant no one was in my flat, so I advertised a room for rent, primarily for the security of having the place occupied all the time. Several people contacted me and I interviewed some of them, deciding eventually to let the room to an Indian couple. I had worked in Chennai, love spicy foods, and thought there might be a perk to this tenancy if they shared curries. Before I notified them, though, I received a strange last-minute call. A young male voice with almost no English was desperate for a place. His name was Jake, and he was with his girlfriend, staying at a backpackers in the city, but wanting to get a better place. They had just arrived in Australia a few days earlier, and the backpackers was grubby, unhealthy and filled with unpleasant occupants. He pleaded with me and the more I spoke with him the more I felt that this person was in danger of being exploited in Australia. I agreed to meet with him, taking an inordinate amount of time to explain directions.

When he arrived he was exhausted and I discovered that he had walked all the way to my place, not knowing how to take a bus and having no transport of his own. He explained that he had eloped with his girlfriend Judy from South Korea, against their parents' wishes. With very little vocabulary it took some thought to realise he was talking about an elopement. Immediately, I let the room to them. I offered to drive with him back to the backpackers, met Judy, collected their suitcases, and helped install them in the master bedroom at my place. I had always intended to take the smaller second bedroom for myself, with the higher quality furniture in the master bedroom making it more attractive to let. Not that this would have made any difference for Jake and Judy.

They were both practising Christians and there was a Korean church in Townsville. Who would have known. I took them regularly to their meetings, waiting in my car reading a book or coming back later to collect them. For a person like myself, with no religious beliefs, I seem to do that a lot. I did the same for friend Louise in San Francisco, waiting outside while she was inside the same church attended by actress Sharon Stone. Driving another friend, my beautiful Filipina friend, Carmen, through Glenorchy in Tasmania. When we passed a Seventh Day Adventist church she said that was the church she belonged to.

'I thought Filipinos were Catholics?'

'I am,' she said, 'but I disagree with much of their teachings and I joined the Seventh Day Adventists since it more closely matches what I believe.'

When I asked her about the attendances she sadly said she had never been to the church while in Tasmania.

'Why not?'

'I don't drive or have a car and none of my friends will take me.'

'Well, we will fix that then!'

Not only did I take her to her meetings, on those occasions I sat inside with her, purely out of curiosity, to discover that the services were just like all the other Christian services I had seen over the years.

'Ours is the only true religion, the only true God. You will be damned if you do not believe.' Fire and brimstone. Not a lot of love and compassion.

A few days later Jake explained that he had searched online for some used bicycles, and was confused about how to go about buying them. Still feeling their vulnerability, I offered to find suitable bikes for them at good prices and then drive my mining truck with them to collect and pay for them. To make sure they were not taken advantage of. Within days, Jake had secured a job as a tiler, with a Korean tradesman in Townsville.

'Did you know this Korean?' I asked.

'No.'

'Do you know anything about laying tiles?'

'No.'

So much for complaints about bludging, whingeing foreigners. This inexperienced young person, in an alien country trying to care for his equally junior girlfriend, against all adversity, had secured accommodation, transport and work. All within days. 'Whingeing' is the Australian equivalent of 'whining' in the USA, and 'bludging' is the worst insult in Australian slang, worse than 'bastard' – it means unrepentant laziness.

There were never any Indian curries, but lots of Korean traditional kimchi. Redolent, salted, fermented, spicy cabbage.

Eventually Jake had saved enough money to buy a cheap car. A friend of theirs had offered them one, and true to form I offered to check it out for them first, then help them to negotiate payment.

When they took ownership of the car, I said to Jake, 'Let's take it for a drive.'

'I cannot drive.'

'What!'

Clearly my next task was to teach him to drive. Up and down the streets of Hermit Park in Townsville.

Some friends from Ethiopia heard about this and suddenly I was teaching instructor in demand, for wives whose husbands could not teach them and teenagers whose fathers could not teach them.

This Korean story was like something out of a romance novel.

Jake was the illegitimate son of a top Korean pop singer and her bass guitarist. Judy was the daughter of a wealthy executive of a Korean electronics company and a qualified nurse back home. Typical Australian regulations would not allow her to practise in Australia, so she secured work as a carer. With their relationship admonished by their parents, or mostly her parents, I think, they had opted to quit Korea and fly to Townsville in Australia.

Jake and Judy prospered and moved out when they found a flat for themselves. A few years later I had left Krucible ingloriously, and was working with companies in Adelaide, South Australia. After a period of no communications from Jake and Jude, I received an email, asking how I was, and letting me know they were in Adelaide.

'You won't believe this, but I am in Adelaide too.'

So we met up again and had a meal together. Next, I heard they had moved to Canberra. Then nothing for almost three years. Friends like this are never far from your mind, and it was satisfying to finally receive an email, saying Judy was now a registered nurse, Jake had his own tiling business, still in Canberra, and they were expecting their second child. I sent a pic from when we first met.

'We were just children,' Judy replied.

Then nothing for another five years, and suddenly a beautiful picture from their wedding, a mature, serious, seriously together couple in classic Korean garb. The picture somewhat like a stilted 19th century family posed portrait, those where you have to hold your smile for five seconds or so, and even sepia-like colouration.

The story of Jake and Judy is one of the gutsiest I have ever encountered.

CHAPTER 25

THAL

Waddamana is set in a beautiful, deep, luscious alpine valley. It has to be to receive the waters down the mountain from the higher penstock to spin the turbines in the power station. Around the penstock is a high, flat, desolate, cold, steppe plain, but Waddamana is paradise. Or so my nan used to say. One of the two favourite songs of hers that I remember is 'Little Green Valley', written by Vern Dalhart and Carson Robinson in 1928, but the version she liked and I heard was by Burl Ives from 1952. Her other favourite song was 'The Three Bells' by the Browns, but originally a Swiss French song and sung by Edith Piaf, which Nan would never have known. Maybe she did and maybe I'm not doing her justice. It's curious how there are always questions one wishes they had asked departed ones. She loved Gracie Fields, for example, who died on almost the same date as Nan. Gracie exemplified the rags to riches trope – the rise from simple beginnings to becoming one of England's most loved singers – but so unpretentious that she would also sing about The 'Biggest Aspidistra in the World', and that absorbed Nan.

You reach Waddamana after a dusty car trip up the southern highlands of Tasmania, climbing the steep, high, misleadingly named Little Den Hill where I missed that turn one time and flew off the side, changing my life forever; you cross the first central plateau which houses the small village of Bothwell with its Scots theme from its Scottish heritage, down the main street past where distinctive short, squat pine trees once lined the road, one for each fallen local soldier of the first world war; then you bounce across the wooden 'kangaroo bridge' over the Clyde River, past the oldest golf course in Australia replete with grazing sheep. After even more climbing to skirt the tortuous foothills leading into Waddamana, you turn the last corner and suddenly, you're staring at those amazing silvery pipelines cascading down the mountain, the waters joining the icy Ouse River. It's almost otherworldly because there's nothing but undeveloped virgin forest until then. Those pipelines were a sign of imminent adventure for us kids – and the loving hugs of Nan. Our 'Little Green Valley' too.

It is perhaps not a coincidence then that the first place my wife Gay and I bought was at an equally remote location called Lonnavale, under the Snowy Range with land bordered by the icy Russel River, that resembled the valley of Waddamana in ways that could not be accidental. Perhaps it was psychological, but something intrinsic in my brain might have attracted me to it.

And it cannot be by accident that the last place I bought, at Wattle Hill, also nestled along a river course called Iron Creek, under Mount Morrison and Block Hill, was another facsimile of Waddamana.

Although most of my time in Europe has been spent in Germany, my first long-term stay in Europe was in France, at Caen near the historical Normandy coastline, with its remnants and memorials of the World War II D-Day landing and history of the Norman Conquest of England.

In France I developed a robot vacuum cleaner for Moulinex, the large appliance manufacturer, whose first product was a coffee bean grinder, a coffee mill, hence the name, Moulin in French meaning mill. Think of *Moulin Rouge* (Red Mill). In honour of my French patron, I named the robot d'Entrecasteaux after an early French explorer of Tasmania, and fascinated my colleagues with the history of the early French exploration of Tasmania including Freycinet's wife hidden on board as a man so she could accompany him. Women were banned from the ships. Several French employees had visited my company in Tasmania, so they were familiar with the place.

Moulinex's European research centre was in Caen, the home of William the Conqueror, near the Bayeux Tapestry, near Le Mont St Michelle, near Paris, near the Britany Alignments, rows of stone menhirs that were old when the pyramids were being constructed, near the Normandy D-Day landings. Packed to the rafters with history. The landing site at Arromanches was particularly impressive, if confronting. During the war it was the largest port in the world, built artificially on the beach from floating pontoons to receive all the troops and equipment for the retaking of Europe from the Nazis. Many pontoons are still there, as are some artillery, and above the beaches there are bomb craters like a pock-marked moon surface and grey decaying concrete German bunkers, all accessible and able to be entered, like an interactive 3D virtual reality show. At times I felt I was in a diorama of destruction.

At the same time that I arrived with some of my staff in Normandy to build the prototype of d'Dentrecasteaux, Sabine, who became a close life-long friend, arrived from her home and work in Solingen in the heart of industrial Germany's Nord Rhine-Westphalia. We were two expatriates, so there was an immediate bond. Moulinex had acquired Krupps, a competing German brand, which unfortunately was soon to become the seed of failure for Moulinex, but at that time some key staff from Krupps were retained and moved to the research centre in Caen, including beautiful Sabine, sophisticated with severely short-cropped hair and a beautiful accent that was more English than the English. After bumping into me at the copy machine she said she intended to marry me, which never happened.

But we are together as friends. Perhaps a more sensible outcome. Since completing the Moulinex project, I have spent all my subsequent visits to Europe as Sabine's guest, initially in Solingen, later in Freising where Joseph Ratzinger was one time archbishop at the Freising Cathedral, later to become known as Pope Benedict XVI, and eventually in Munich where her flat is adjacent to the beautiful, incongruously named English Garden, with the Isar River meandering through, and close to the universities where the structure of the universe was unravelled. Close also to the birth of German fascism.

Solingen is throwing distance from Dusseldorf, Bonn and Cologne, three beautiful historic German cities along this final path of the Rhine River before it finally escapes Germany for Holland where it magically becomes the Waal, having travelled all the way through the country south from Basel in Switzerland. Cologne has its church, Dusseldorf has its fashion industry and Bonn was the political capital of Germany until the reunification of Berlin on the collapse of the Berlin Wall. Solingen is where the famous highest quality knives and scissors come from. I have a cyclorama I made illustrating a walkthrough I had of the stages of manufacturing scissors in a pre-World War II factory, now a museum, in Solingen. I visited and collected the bits for the display off the floor. For a technologist it was incredible to me that there was not a single electronic device in the factory, not even electromechanical; all presses, stampers, cutters, moulds, foundries, burnishers, polishers, gold platers. Just brute force.

In Solingen it was also frustrating to see that the city still had all its trams and trolley buses, maddening because in my home city of Hobart they were ripped out by ignorant, shortsighted officials. They're still doing the same today with rail, anything that is not heavy road transport. Philistines. Saddening.

The industrial might of this area was so crucial to Germany military operations during World War II that it was decided by England to try to destroy the Endersee and Scorpe dams across tributaries of the Rhine, the reserves of which fed the hydroelectric power stations supplying factories that were making arms and ammunition. Supplying factories like the scissors one, which had been transformed to make weapons. Access to the dams by plane seemed impossible because of the difficult terrain, steep approach requiring bombers to nosedive at the last minute, and the need for the planes to somehow get so far into Germany's heartland. For similar reasons, the power station at Waddamana was a sensitive site, protected in case it was the target of bombers during the war. I remember concrete bunkers at the right-hand side of the road leading into the village which we used to crawl into and play soldiers. The solution to the German dilemma was Operation Chastise, with the well-known bouncing bombs of Barnes Wallis, better known as The Dambusters, indelibly imprinted in everyone my age by the admittedly glorified movie. Glorified because instead of the single successful run in the movie, over 50 air crew were killed and 8 aircraft lost in multiple raids, with almost 2000 Germans dead in the resulting flooding of the Ruhr and Eder Valleys.

One day in the local Solingen newspaper on the anniversary of the Dambusters, images were published of the destroyed dams, photographed by Germans from the German side, showing the damage as the Dambuster air crews would never have seen. It was fascinating to get a feel for the reaction from the other side. The associated article was admiring of the ingenuity of Barnes Wallis. It had a similar effect on me as the D-Day landing museum in Caen had had, with its double movie screen showing simultaneous films of the landing taken by both the Allies and the Axis; the adrenaline and fear and anticipation of the troops landing and the simultaneous sleepy soldiers at Normandy waking up and realising they were being invaded. That alone gave me the strongest impression of the war, more than anything else I have read or seen.

Not all valleys are utopian. I lived and worked for four years in Pittsburgh. It is a valley at the confluence of three large transport rivers: the Ohio, Allegheny and Monongahela. The local sports stadium is called Three Rivers. The local park is called Three Rivers. The sides of the valley are steep and the only way up the inclines at times is via funicular tram lines. Pittsburgh was industrial writ large and the history of the city is not pretty. Because it was the centre of coal, iron and steel, and the need for factories to be along the river flats near the rivers so that raw resources and finished product could be unloaded from or loaded into barges, the land was a premium for the industrial barons of the day. Carnegie, Mellon, Frick, Morgan, Swarb, Westinghouse. Not just iron and steel, but other manufacturers, Heinz Soups, for instance; America's first steamboat, one of the first locomotives, each were built there. Factories, smoke, dirt, soot, pollution, coal dust, all crowded the valuable river flats. The wealthy lived in green undulating hills away from the mess while workers were relegated to the poorer land on the steep hillsides within breathing distance of the mills. Pittsburgh was also the site of massive worker riots and ruthless oppression, with companies using private mercenary armies to stop strikes and union formation, by shooting workers when necessary. Texas has no monopoly on guns.

When the skyscrapers were being constructed in New York and Chicago, the steel girders arriving by train from the foundries of Pittsburgh were in such immediate demand that they were sometimes still hot. In the 1970s, though, with the downturn in heavy industry, decreased raw material needs and primary industry wealth decline, things changed. The factories and industrial sites along the three rivers were abandoned, cleaned up, and the land vacated. Instead of the city selling off this valuable land, which would have restricted access for the city population to the potential presented by river flats, it was turned into community space, with parks, promenades, sports grounds, museums, cafes and restaurants. Industrial warehouses were turned into some of the first entrepreneurial hubs and artists studios, or cheap inner city housing. It was visionary. Unlike the privatisation and sequestering of prime public land in Tasmania for vested interests.

The result was the city converting from the rusty image of Iron City to a clean, vibrant, progressive city attracting academics, clean industries, movie making, teaching hospitals, advanced technology centres and high quality of life. Universities expanded, cultural centres like Heinz Hall flourished. Pittsburgh today is a model of urban renewal that cities everywhere would benefit from studying.

A world away, near Solingen, is a similar steep-sided valley along the Wupper River. The city winds along the river flats, and the banks of the hills quickly rise to limit the extent of the city. Its nickname is the Linear City. About 120 years ago, with the pressure of transport in the growing city, they needed and wanted to add a public rail line, but there was no land available. So, with typical German innovation, they decided to build the rail along the centre of the river. The only way of doing so was to suspend it high above the water. To add to the ingenuity, they settled on a monorail design, not just any old monorail but a suspended one, and as if that is not enough, it is electric. So, the Schwebebahn or 'floating rail' was built, the oldest of its kind in the world. It carries over 80,000 passengers each day, including me a few times, along a 10-kilometre course with the best views of the Linear City and the Wupper River, the most famous passenger being Tuffi, a circus elephant who not surprisingly got scared during a promotion on the line and survived a jump into the river.

Nearby is another river I have visited, the Dussel. The city of Dusseldorf, meaning Dussel Ville, sits on this river where it joins the Rhein, but a bit further upstream is yet another steep valley called Neander. It used to be the site of intensive limestone quarrying for mortar and plaster, then one day in 1856 in the valley mines and caves, bones of a prehistoric person were dug up, clearly not quite Homo sapiens, yet not an older pre-human skeleton. Naturally it came to be named after the place where it was found in the Neander Valley. The German word for valley is Thal and the ancient person became known as Neanderthal. I've been there where the limestone quarries are long gone, but there is a fascinating museum showing these people, with brains larger than ours, who lived alongside us until just a few thousand years ago, and might still be lurking among us.

CHAPTER 26

HOME AND AWAY

Dad was hard to live with; just ask his two wives and one girlfriend. Or ask any of his kids. It got worse as he got older and as his failed life set in.

One of the problems was that both Mum and Dad were as physically strong as each other. Dad was about 5 foot 8 inches and Mum was about five foot one. And they would hit to hurt. They were equally confrontational and uncompromising. We did not know that waking up to see Mum with black eyes, or Dad with cuts and bruises, was not normal. Many of our parents' friends were from the same socioeconomic group, heavy drinking, heavy partying, heavy fraternising, so we saw that in other families too. Later, when they finally successfully parted ways and it no longer mattered, Mum and Dad were best friends.

After Dad and Mum split up, they both had other partners, but Mum settled with a great man named Arthur until she eventually resolved to live alone and focus on family. Dad settled for a much younger woman called Sandra, the sister of June Willmott, whose family we grew up with. Although it was another passionate but unstable and disastrous affair, it was nevertheless long-lasting, so Sandra was a large part of our lives through our teens. That relationship, though, was as traumatic as his marriage, so it was like nothing had changed for us kids. After Sandra he married Mary, which was short-lived, and they were estranged when he died. It is probable through that time that Dad was suffering personality changes caused by his brain tumour, so Mary probably did the best she could to live with a difficult man becoming more difficult. I liked both Sandra and Mary very much and still visit Sandra to this day.

My youngest sister Carol never lived with Dad, always with Mum. Sister Olannah also opted to spend time with Mum, and she and Carol were close in their formative years. Sister Sally got pregnant and left home, able to fend and support herself from the beginning.

My other sister Dianna stayed with Dad the longest. They taunted each other, so her stories are legendary. She is the family raconteur extraordinaire. One morning over breakfast, Di, as we call her, was discussing the wood heater box they had in the living room. Dad must have been in one of his rare, meaning sober, moods. Di said she could not understand how the glass of a fire box never shattered, how it stood up to the heat.

She said, 'Dad, that glass must be something really special.'

'Of course it is, it is hardened and tempered to withstand the heat.'

'But other glass could never withstand the heat, could it?

'No. It is special silica glass like they make Pyrex glassware from,' he continued. 'Also the frame the glass is in has room for expansion, otherwise it would still shatter.'

'That is some glass!'

'Yep.'

'Glass like that would be really expensive, wouldn't it?

'Yep.' Pause, more silence … then, 'You didn't?'

'Yep.'

'How?'

'I could not get a log of wood all the way in and I tried to jam the door closed.' After a long pause of absolute quiet, Di said, 'It made a huge bang noise, like an explosion.'

'I need a drink.'

Dianna never hesitated to make herself the butt of any story. One day she took her car to the mechanic. 'It starts very well, but after a while it shudders and stalls.'

'That's okay. We'll check it out and find out what is going on. Leave it with us so we can take it for a test drive.'

When she went back the next day: 'How did you go?'

'We could find nothing wrong with it. Two of our mechanics took it out and it runs perfectly.'

'But it does it all the time with me, every day.'

'Can you come with me, then, and you drive, show me how it shudders?'

So Di jumped in and pulled out a knob on the front dash to hang her handbag and took off.'

'Why did you do that?'

'What?'

'Pull out the choke.'

'What choke?'

'That knob there.'

'Isn't that a hook for handbags?

'What?'

'What!'

'I think I'm having a stroke!'

Then she tells of the first time as a young adult she went to the dentist. During the procedure the dentist placed a little paper cup of pink fluid down for her, then went into the other room to prepare something. Di picked up the cup and drank it. The dentist came back into the room, stared at the tray looking confused, then put another cup of fluid down.

'I could have sworn I put one down already.'

Di said, 'Do I have to drink that one too?'

Stunned silence, like it was taking a long time to compute, then, 'What? No! It is to rinse with.'

'Thank goodness! It tastes like shit.'

These anecdotes belie her amazing skills, mostly in social settings; she also has a talent for acting and a greater talent for presenting in public, for holding court. I was surprised when at our mother's funeral, she stood at the podium and delivered part of the service. I was shocked and astounded to hear her beautiful, resonant, confident voice. If Morgan Freeman had a female equivalent, it was my sister Dianna. Then a few years later at our sister Sally's funeral, as I was struggling to compose myself as I read my eulogy for her, it was Dianna who rose from the crowd and came to stand beside me and hold me up.

My younger brother Toni did not last as long at home. In fact, he was the first to leave. Not once but many times. From about the age of 14 Toni ran away from home on a regular basis. Each time he would be brought back by police. Each time he got further and further away. As difficult as Mum was, she was much easier to live with and Toni eventually made it to her place and stayed there with her at Ocean Grove in Victoria. On one occasion, he was caught by police in the Queensland resort area of the Gold Coast, breaking into a supermarket, starving for something to eat. Dad received a call from the interstate police about 2000 miles away, saying they had his son.

'Keep him,' was Dad's reply.

'What?'

'Keep him. I'm sick of it; he is always running away.'

'We can't keep him. We have to bring him back. You are his legal parent.'

'I know. I just wish you could keep him.'

The police arranged to fly him back with a police escort on the plane to protect him but also to ensure he got back. At the transit stop in Melbourne, Toni escaped. He ran away from the police. They searched for him and never found him. They assumed he would head to his mother's rather than his father's place, but instead he went north again and ended up staying at Kings Cross in Sydney. He said it was a long time before he realised that the array of beautiful bevvies living around his area were prostitutes, Kings Cross being the red light district of Sydney, and today the site of the Mardi Gras parades. Toni said he was broke every payday.

Toni was unashamedly sneaky. At a slightly earlier age, he was already running away, but not from the house, within the house. Often no one knew where Toni was. No less shrewd, Dad discovered a trapdoor in our bedroom leading through the floor to under the house. Toni had found it and fabricated a warm, cosy cubby hole down there, not high enough to stand up, lined with blankets for walls, a light, and food. Cans and cans of food. It was like he was practising for the real thing a year or so later. None of us knew about it, including me, despite the fact that I shared the same bedroom with him and the trapdoor was under my bed.

There was only one time I ran away. I made it to the first country town north out of Hobart. It was dark and cold, so I stopped and lit a fire off the side of the road in a field. I was just settling down when the land owner came by and asked what I was up to. Embarrassed to say I was only a few miles on a trek north, I said I was hitching to Hobart but was tired. He drove me to his barn and said I could sleep there for the night. I didn't sleep all night because something, snakes or rats, crawled over me and I was too scared to move in case it bit me. Next morning I had had enough and hitched back home, arriving just before anyone woke up, and started making breakfast in the kitchen.

Dad was first to come out and he said, 'Gee, you're up early. I'll have some tea thanks.'

Running away was Toni's trick, not mine. Throughout his adolescence, he was Spitfire mad and all he wanted was to be a pilot, which was an impossible dream, of course. Toni died while I was writing this memoir, but for someone with his auspicious start, Toni had a stable working life, with just a handful of very long-term jobs. Always liked by bosses and colleagues. One year he represented his town when it won the state's Tidy Town competition, borrowing my suit and tie for the public celebration and awards ceremony. He lived in this town, Bothwell, around the corner from where our great grandmother Fanny Branch lived, a block from the business I was managing the night I had my life-changing car accident, for 30 years, working for the regional council. Toni, like his sisters, was innately intelligent. His kids think he is the smartest person they know and to see him at his crossword puzzles you'd believe it. He was a dark horse. His job for a long time was to manage the town's water quality control, which meant weekly trips across a wide, dislocated mountainous area, collecting water samples from all the streams and water courses forming the drinking water collection basin, then transporting the samples to the assay labs in Hobart, and back to Bothwell to fluoridate the water supplies after testing and calculating dosages.

When Toni was conscripted in the bizarre and macabre raffle held by the Australian Federal Government to call up soldiers for the Vietnam War, I sent a message to him referring to the Righteous Brothers' song 'He's My Brother', a memory he often related in appreciation to strangers. For my recent birthday, he invited me to lunch and as a special dispensation he invited me to sing and play my guitar, something he normally loathes. Now that he is gone this kind act gives me déjà vu about my dad's telephone call to me in Dallas before he died. I suspect Toni knew his end was nigh.

Mim and I were in Dallas, working at Commodore Computers, when we received a Saturday morning call from a stranger with a very broad Australian accent. Our accents had started to modify with the southern Texan drawl around us, so an authentic Aussie accent stood out.

'It's me. I'm here.'

'Who?'

'Steve.'

'Steve who?'

'Steve Powell.'

'Steve Powell who?'

'Steve. With Hartz.'

'Hartz who?'

'You know,' with increasing frustration, 'Toni's boss.'

Steve was on an international marketing trip. He ran some food and beverage businesses in Tasmania, and regularly toured the globe looking for product and marketing trends. When he told Toni he was going through America, Toni said that he should call on us, and we would host him for a visit. The trouble was that Toni forgot to tell us.

We raced to the airport to pick him up, losing our entry ticket in the rush so we had to pay the full parking fee to get out, and he stayed with us for several days, experiencing Texas and Dallas and Billy Bob's, and at the end of his trip, Galveston, by mutual consent.

Steve had short thinning hair, was driven, energetic, ruddy faced, average sized … he'd had an average education but had become a hugely successful self-made businessman, very generous, struggling to keep his weight under control, an emerging epicure. Although Mim and I had been in Texas maybe a year already, work was intense and we had not toured a lot ourselves. We had been to Port Arthur (not the infamous Tasmanian one with its mass murders, but the industrial oil port on the coast of the Gulf of Mexico), for a wedding in Beaumont, the town next to Port Arthur, almost into Louisiana. One of my key engineers was getting married, and we were invited to the celebrations, travelling down through the surprisingly verdant pine-forested eastern Texas, unlike the rest of Texas with the desert out west, and discovered to our bemusement that it was a Southern Baptist wedding. We were the only ones to present a gift, and the reception was in a pew-furnished room adjacent to the church, where we took a slice of cake and an orange juice then left.

But Galveston was an interesting idea. We knew the Glen Campbell country song of the same name, and the area had been newsworthy because of Hurricane Alicia destroying the coastal settlements in 1983, the year we arrived in Dallas. So, we set off for a weekend trip to Galveston with Steve. South past Waco, famous as the home of Dr Pepper, a soft drink released a year earlier than Coca-Cola, but was not yet famous for its mass murders of the Branch Davidians, (no relation), through sprawling Houston, to the Gulf of Mexico, across the bridge to the island strip that is Galveston. Waco looks a bit like 'wacko' and when I was DJing on that radio station in Tasmania I started a program that I called 'Waco West" of non-mainstream country music: Cajun, zydeco, swamp-rock, rockabilly.

Dinner that night, after booking into a motel, was at Gaido's, a famous family fish restaurant memorable for their sign which read, 'Not a Chain, Never Will Be', with spectacular views across the sundrenched waterfront. Steve saw Australian Fosters beer on the menu and ordered one; he was shocked when it arrived almost double the size of an Australian beer can. His marketing eyes lit up. As the typical Texas-sized beer arrived at our table, a group of people sitting across the restaurant rose and moved towards us, like an attack in a Chinese movie. A rush of Australian accents came at us, with the table explaining that they assumed we must be Australians when they saw the Fosters.

Back at the motel we checked into our rooms and arranged to meet in the lobby next morning for coffee and to check out. But next morning was when everything went wrong and got really terrifying.

Steve was late. We waited in the lobby for a long time, not wanting to rush him, but starting to fret. Eventually we asked reception to call Steve Powell in his room. They checked the register and said there was no one in that room. We argued that was impossible because we saw him check in last night at the same time as us. We saw him go into the room. They said no and showed us the register with no occupant for that room. They went to the room and came back saying it was empty and not used. We suggested they just check for a guest named Steve Powell, he must be in another room. They went through everything and there was no Steve Powell. By now it was several hours later and we were panicking, wondering if something perfidious had happened to him. This was not a joke, and this was Texas after all. On a weekend road trip to El Paso one time, Mim and I were delayed at a roadside crane that was lifting a car that had crashed off the steep embankment back onto the highway. We asked the trooper what had happened and they said someone shot at them because they did not have Texas licence plates. We were driving Commodore's company car from Philadelphia with its Pennsylvania plates. Unbelievable things happened in this part of the world.

Just as we were ready to call the police, thinking Steve had been abducted or worse, into the room he came, bleary-eyed with a hangover that only giant Fosters can give you. It turned out that in the middle of the night he called reception to complain about crickets in his air conditioning keeping him awake. This coincided with the changeover to a night manager, so the documentation got missed. They moved him to a different room, but never recorded it and cleared the old room so no one would be checked into it until the insect exterminators had been through it. He had no idea of the problem that had been caused. Having had little sleep, he turned off any morning alarm and simply slept through. It reminds me of another Texan, Buddy Holly, not from Galveston but from Lubbock, and his band The Crickets, named because of a similar problem when their first recordings in their garage had the unmistakable sound of chirping crickets in the background.

As we drove along the Galveston coast, which is mostly low elevation sand dunes, I felt that the dilemma we'd had with Steve was reminiscent of that Glen Campbell Galveston song. The songwriter, Jimmy Webb, said his words were about someone caught up in something he did not understand.

The sobering remnants of Hurricane Alicia were everywhere, with few standing structures, just piles of rubble. Damaged boats collected like foamy flotsam along the beaches, seagulls were absent, a few people absently sifted through the debris, clearly wishing they were somewhere else. The gulf was still as a pond, the colour washed away like a faded favourite t-shirt, quiet as a silent movie, a perjury to the destruction it had caused.

We were in our mid-teens about to experience the scene of another destruction. My brother and I were shocked to see our names come up in the middle of a halted movie at the local cinema. A hand written emergency note zoomed onto the giant screen. It told us urgently to go to the office in the front of the theatre. It was surreal, we were in disbelief, and when we walked, embarrassed, out of the dark room where the movie had resumed, we were confronted by police. They said that our father had been in a serious car accident and was in hospital in ICU. We raced to the hospital but were unable to see him, and went home. The next day his totalled car was hauled in to our driveway. That was the one Pa Branch advised me the best way to panel beat was on a sandbag, but the damage was so extreme the car was never rebuilt. Nor was Dad's life. This was the end game of a life of bad choices. Although he continued, got married again, had a regular job driving a Australian Post delivery van, he was always a shadow of his former self. In some ways he stabilised, softened, but a weaker version of a hard man is still a hard man.

Until this time, we had been moving from house to house, Dad's houses, the fixer-uppers he restored. But eventually he resorted to renting as his jobs became short-lived and our own houses were mortgaged then sold as the money dried up. The state government provided affordable housing and we applied. Dad was in hospital again and I forget the details. My boss at the time was Ron Richards, head scientist for horticulture for the state government, and it was me and Ron who completed the submission. When the application was successful Ron used his company truck to move us out to the new house at Erskine Street in Claremont, the last house Dad ever lived in. Ron was an early father figure to me, like each of my later bosses and some more mature colleagues. These people took me under their wing and helped me to break the cycle of my family history. One time Ron had loaned me his set of socket spanners to work on my little Morris Minor, my first car. I kept them in the unlocked boot of my car. Ron secretly took them out of the boot of my car one night. The next day he came by and asked for them back. He relished the surprised shock that came on my face when I opened the boot to find them missing. He told me he was teaching me a lesson on caring for things, keeping things secure. When Dad had recovered enough to leave hospital, we took him to his new house, all moved in, courtesy of Ron Richards.

Elsewhere I mentioned that this is not how Dad started; how in the beginning, he was an ideal father, but the forces were against him. His inner demons overwhelmed him and there was no support. Actually, we all supported him, but there was no way out other than to try to keep him safe from himself. He died very young, 53, from an inoperable oversized brain tumour. He must have had it for a couple of years and never knew. He was mostly hungover every day by then, so he would not have known the headaches were from a disease. It explained his extreme painkiller taking, chewing on paracetamol like candy, and his lack of coordination that we had put down to being drunk. We were probably unfair to him; perhaps his drinking was to mask the pain he was in.

I also remember his industry, the ingenuity, the dreams and aspirations and attempts. Once, not content with the commercially available clay bricks, at 41 Goulbourn Street, he and his friends made handmade bricks from concrete, using a mould that had an interesting pattern across the face, to construct the front fence. The houses he restored, the playground equipment he made for us, the games he played, the involving of us in his activities, sharing the music he loved, and dancing and telling jokes. His love of the country, and hunting, and of cooking. He liked women and women liked him, but he was also a man's man. His male friends looked up to him. They were all giant personalities, but he was the leader. He would have benefitted from his own 'Ron Richards' to help him break his cycle, but there was no one.

Initially I tried to emulate him, but there was something different in me. I recall sitting at primary school at playtime with a girl on the wooden benches beside the sports oval. I think her name was June. She was clearly a lonely and poor girl, with a hand-me-down ragged school uniform, and was made fun of by many. Taunts like 'She stinks!' were common. I sat with her to give her company, and to make a statement to the others, and to her maybe, although I would have been too immature to understand even my own motivation. I never felt peer pressure, never felt it necessary to shun her like the others did.

Russell Rodman had a reputation as the fastest runner in the same school. We were all under 12 years of age, playing chasings at lunch time around the same vast oval. During one game when he was 'it' I was racing to catch Russell and realised I was closing in on him, so I slowed down to not ruin his reputation.

But I could react violently to injustice, whether to myself or others. Thugs living adjacent to my cousin Serena picked on her. They were notorious and dangerous. Actually, small-time criminals known to the police. Playing at their own game, I confronted them with threats that they knew nothing about me and what I was capable of. I dropped into the conversation that near where I lived on Nugent Road underworld bodies had been found. Like many bullies, they caved in when it was not a vulnerable female they were attacking. Bully matched by bluff.

Something must have been troubling me, though. I had troubling dreams. I'm not a fan of Freud. I think he was wrong about everything. So his definition of dreams as he described them, wish fulfilment, does not apply as far as I am concerned. The road into Waddamana was a steep, winding, mountainous, single lane dirt track. The drop-off over the edges was terrifying, I had recurring dreams of going over the edge, clearly not a wish I wanted fulfilled. Another recurring dream I recall clearly involved me being on the bank of the river at the local annual Royal Hobart Regatta. Climbing up over rows of terraced walls, only low walls, but being paralysed and unable to scale them, crying out to Mum, who was getting away from me further up the hill towards the Domain and Cenotaph. Today I laugh at the joke about a child lost at the fair, picked up by a police officer who asks. 'What's your Mummy like?' and the kid replies, 'Beer and men.' But a realistic dream of losing your mum is not funny. Then there were my astral flying dreams, floating just out of reach of kids on the street who were grabbing up at me. Lying half asleep at night I'd will myself up into the air, and sure enough with enough mental effort I would start to levitate. In my dreams.

There's a recurring theme in my life of looking out for the underdog, which pervades even my corporate turnaround work, but also something of undefined desperation or disappointment that the world is not fair. Not designed to be. Necessarily unfair, else we would not have evolved.

CHAPTER 27

WHY AREN'T YOU IN COURT?

There is a famous Australian broadcaster called John Laws. Controversial, abrupt, intelligent, interesting, successful. He liked country music, which is the only thing I liked about him, that and a poem I heard him recite once. In his cavernous velvety voice he talked about the rules of the toilet, reciting some of his pet hates and pet likes; I forget the details and the rhyme, but I loved the punchline ending when he said that these are the laws of the john, or, the john laws.

I know a lot of Johns, the people kind, not the toilets. Well, the toilets too, from the drop pit at South Arm, to the standing only squat toilets in Asia, to the almost open to the public toilets in Amsterdam, to the bidets and hose toilets, to the handwashing basins in Malaysia, to individuals who sit or stand on the seat, to the free use of male or female toilets by males or females in France, to families where toilet doors are left open, to our family which keeps it very private.

But the people type. There is a software technician John, who believes that when Europeans arrived in Australia, they had a war with the inhabitants and won, and who believed that women became excited by the musky body odour of men. There is Jon Jarvik, my scientist mate in Pittsburgh, or equally as brilliant, my scientist mate John Reid in Hobart. One a geneticist, the other a cosmologist, their minds delving into the microscopic and the macroscopic. Then there is John the lawyer.

Over the years John Avery represented probably every one of my close relatives. It says more about my close relatives than about John. I was introduced to him by Dad. He said John had done a good job over a traffic matter for him, and while I forget what I needed him for, because I have no memory of every needing a lawyer for a traffic matter, that is probably what it was. Likewise, I forget what became of that matter, but it must have been successful because as a consequence, the time I really needed a lawyer I called on him, and every time since.

John is one of those larger-than-life lawyers that you expect to see alongside the celebrities in Hollywood. He is wasted in a small town like Hobart. A hidden personality. He is a patron of the arts, dressed unlike a lawyer, even in court, as if he is fresh off the golf course with plaid trousers and casual polo shirt. I know lots of lawyers. Most companies I have turned around have their own lawyers, and to a one they are uncolourful.

John got into trouble, but he is my mate and I am loyal to my mates. Medium height, a lover of good food, high forehead and thin hair, robust body, handsome face, confidence writ large, boldness beyond description. He's had the wind knocked out of his sails but the sails are set again.

John helped me when I needed court representation so I could retain legal custody of my tiny niece and nephew whom I had taken legal guardianship of when they were abandoned by their father, after their mother died. That was a big matter, nothing trivial like a traffic fine, because the father fought me in court, seemingly angrier about my actions than concerned with looking after his children. That proved to be true, because after the custody case, he again went absent for many years. John resolved that matter; he had the wit of Whitlam. The smug and arrogant father, in the court witness box responding to a question with, 'Do you think I am stupid?' received the instant retort, 'We don't have time to debate that now, but it might be useful later.'

I was in Texas, in a discussion about gun control laws and mass shootings in the USA, proudly, and in retrospect pompously, saying how these did not happen in Australia and certainly not in my home state of Tasmania, when the news came on about Australia's worst mass shooting in my beloved Tasmania. Thirty-five people shot and killed by a lone madman at a famous historical town called Port Arthur, already the site of an infamous penal prison, Australia's Devil's Island, and now much more infamous. Actually, it was the country's greatest non-Indigenous mass shooting – there had been worse colonisation massacres. John reluctantly undertook the task of representing the gunman, receiving both criticism for representing a mad murderer and praise for upholding everyone's right to a fair trial. The gunman, who like many insane did not look crazy, had firmly intended to plead not guilty, which would have cost the state a fortune for the trial. John's amazing contribution was to convince the killer to plead guilty. John became Tasmania's, maybe Australia's, most famous lawyer and world renowned. Emerging from a small pond to a big pond and John has compared that with my career.

Another time John dealt successfully with a labour relations matter when I sacked one of my managers who had stolen from me but who then, like many who are caught out, decided to sue me for unpaid wages.

That Monday was one of those out-of-control days when everything went wrong after the telephone call. The call came as I was settling into a fresh brew of tea and a delicate conversation with Mum about her progress with her long-suffering lung cancer problems. The telephone call was an unwelcome interruption.

'Where are you?' It was my partner Mim.

'I'm at Mum's. What's the matter?'

'John Avery's been calling; he is desperately trying to find you.'

'What does he want? Did you tell him where I am?'

'I don't know what he wants. He didn't say, and I didn't know where you were. I've been ringing around everywhere.'

'That's okay; I'll call him now.'

Mum was looking at me, respecting my privacy but clearly intrigued about the call.

'It was Mim,' I said. 'She said John Avery is searching for me.' Suddenly we both knew what it was all about.

Mim, my Mum and I had a strong connection with John because he'd helped us to gain custody of my niece and nephew. He'd also helped out with some other issues for me, so I knew from personal experience that he was capable across a wide spectrum of legal issues.

'Where are you?' John said in a pleading, desperate voice when I called him a few minutes later, a voice I had never heard him use before. 'Why aren't you here in court?'

'I didn't know I was supposed to be in court.'

'Didn't I tell you?' He said.

'I haven't heard from you about anything!'

'Oh! Damn! That's right. I forgot to let you know,' with a sense of memory creeping back into his voice. 'I'm in court now, they are here, and the judge is angry. You need to get here immediately. It's the Bothwell Shop matter. Where your ex-manager has sued you.'

I was in jeans and t-shirt; not the best image for court. Not a hearing, the main action. I told John I needed to race home, about 45 minutes away, to change to a suit, and then it would take me another 30 minutes to get back to the city and the courtroom. I'd be as quick as I could. As I sped through the streets, I was thinking about frustrated judges and wondered whether the next time I needed John it might be for contempt of court, and it might also be for another speeding ticket.

When my sister Carol died and I decided to take custody of her children, John was amazing. It is a significant decision to take over responsibility for someone's kids, and it was not made lightly. I was in the early halcyon days of an international entrepreneurial career, taking a technology startup company to Silicon Valley, and it meant I was not going to always be available. My partner Mim and I decided to ask my mother to move in with us, and help both of us share the time needed for the two infants. That had been more than a decade earlier, and the kids had grown up and started lives of their own.

When I met John outside the courtroom, he pulled me aside, no apology for the mix-up, and said that while waiting for me he had had time to discuss the case with them, and that they had agreed to a settlement of $40k, reduced from what they had sued me for.

I couldn't believe what John had just said. 'No way,' I quickly answered.

'It's a good deal.'

'They are not getting anything.'

'They have the best lawyer in these matters. it's his speciality and he has not lost a case. You should take it.'

'No!'

I was surprised at the force and disappointment in my voice. I was fuming that they had the hide to expect me to pay something, and disappointed with John to think for a second that I'd settle with them.

'So what do you want?' John asked me.

'I want to give them nothing and to have their claim dismissed, and to have them pay my costs.'

'This court does not award costs.'

'Oh!'

'The risk is high and if you lose you will have to pay all of their claim as well as my fees.'

'But they screwed me over. I am the one who should be suing them. I will not give them a single cent.'

'Are you sure? The judge is waiting in there and we have to go in now and tell him what we intend to do. I couldn't advise him until I talked with you.'

'I'm sure.'

'It will cost a couple of days legal expenses, you know.'

'I know, but I'd rather pay you than them.'

'Okay, let's do it.'

We entered the courtroom, which is always intimidating for the uninitiated, with the judge glaring down at me, trying to be patient, but clearly not.

'Where have you been?' he asked me directly.

'I have been with my mother,' I answered meekly. 'She is gravely ill and I went there forgetting about the date for this hearing,' I lied.

John interrupted, telling the judge that we intended to proceed. I scanned the courtroom for the first time, to see the plaintiff and their lawyer on the other side. I searched their faces, trying to read their reaction to the fact that we had not followed through on the settlement offer, which they clearly thought was going to happen, perhaps because John had led them to believe I would accept.

John requested a continuance to prepare, knowing it was unlikely to be allowed given the mess at the start of the day. Sure enough, the judge ordered us to start. The plaintiffs had to make their case first, so that gave us breathing space, and I could tell from John's taking furious notes that he was not prepared to defend the matter. That was confusing.

But one thing John can do is think on his feet, and because I actually had a valid defence, after two days we won, and their claim was dismissed.

'So, who did the banking?' John asked the ex-manager in the witness box.

'I did.'

'Who checked the money you collected from the till, and the deposits into the bank other than you?'

'No one.'

'So, you were trusted.'

'Yes.'

'And who ordered the goods for the shop?'

'I did.'

'Everything? I mean, were you told what to buy, what to choose for the store?'

'No, I chose and bought everything.'

'Who was your supervisor?'

'No one. I saw the owner now and then, but I was not supervised.'

'And where did you live? This is a general store in a remote country town, isn't it?'

'I had a flat at the top of the shop.'

'And how much rent did you pay for the flat?'

'Nothing, it was included in the deal.'

'And did you pay for personal food from the shop?'

'No, I was allowed to use anything for my personal use.'

John was unrelenting. 'So, let me get this straight. You were in charge of stocking the shop, you were in control of all monies including banking, you received a salary, and you had free rent and food?'

'Yes.'

'Does this sound like the work conditions, perks, that a supervised employee would have?'

'Well, yes, maybe.'

'You don't appear sure. Here in this file I have the description of all retail awards for shop workers, and none of them include a position where the worker has unsupervised access and control of the money and the bank account, none include free accommodation, and none have free rein over what to spend the money on for supplies for the business.'

'But I was responsible for cleaning the floors and such low tasks like a low level worker.'

'What was your title?'

'I was the manager. At least they called me the manager, but really, I was just a worker.'

'Did you refer to yourself as the manager when you spoke with other people?'

'How do you mean?'

'Well, for instance, who did the people at the bank think you were, bringing all that money in each week?'

'Just a worker.'

'Is this your signature on this bank deposit slip?'

'Yes.'

'And is that your position title under your signature, in your handwriting?'

'Yes, I think so.'

'And what title did you give the bank on this form?'

'I don't recall.'

'Please read from the bank form.'

'Manager.'

'So, to correct your testimony, what is your position title at the shop?'

'The shop manager, I think. You're way ahead of me!'

'You bet.'

I looked at John, not a hint of the irony,

'Did you hire anyone to do the cleaning up?' said John.

'No.'

Not the answer he would have preferred. 'Could you have done so if you wanted to?'

'I don't know. Perhaps not.'

'But you had total control over all the money and what to spend it on.'

'But that was for shop products.'

Critical to depart a line of questioning if it is not leading where you want it to go.

'Were there times the shop was particularly busy?'

'Yes, through holidays, the fishing season, we had many people passing through the town.'

'Did you have anyone help you?'

A delay as she realised the trap she had fallen into.

'Please answer the question.'

'Yes, I had a friend help me.'

'And did you hire her and pay her?'

'Yes.'

'Did you ask permission to hire your friend to help you when the shop was extra busy?'

'No.'

Got there. Now checking his notes, on to the next point. I looked over at the plaintiff's lawyer, who was now slumped in his seat.

'Did you use a car?'

'Yes, but it was my car.'

'And did you buy petrol and pay for maintenance for the car, new tyres and so on, with shop money?'

'Yes. I used the car for work.'

'But tyres and petrol are not products for the shop, they are ancillary things, are they not?'

'I guess so.'

'So, you could decide what and when to spend the money on other things?'

'Well, yes, when you put it like that.'

'Can I put it to you that in fact you were not only the manager of the business but the top manager? That you had an extremely generous position, where you lived in a free flat, fed yourself from the shop, ran your car at the shop's expense, were completely unsupervised and trusted with collecting, spending and banking the money. A position that you took advantage of, and once you were found out and fired, you decided to try to get back at my client through the Fair Work laws, like a sore loser. You are simply a vengeful one at that.'

Since it was a Fair Work matter, no costs were awarded and I had to pay John. As we departed the courthouse at the end of the proceedings, John told me the size of his bill. I don't have that much, I advised him.

'How much do you have?' he asked.

'About half that,' I said.

'That's okay, just pay that to the office and that will be enough.'

Apart from his wry sense of humour, another intriguing side to John was his sense of fairness, lack of avarice; he was accommodating to his regular clients, as I had become. Perhaps he was making enough from all of my other family members who I had passed on to him.

'Does that include the parking fine?' I queried with a smirk.

This was not about a ticket of mine. On the second day in court I first met John early at his office, because that was the day for the beginning of our side of the case. We went over the worker awards to prepare to prove to the court that the plaintiff was a manager and not a general worker governed by an award under any regulation or legislation. We drove to court together in John's bright red convertible Mercedes, and he pulled up at a no-parking spot right outside the courts. When I asked him about getting booked, he matter-of-factly said that any fine would go onto my bill.

'That will include everything,' he concluded.

At the end of the first day, Mim asked me how it was going. I explained how I was surprised John had thought I'd be happy with a negotiated settlement in their favour.

'I think there is something wrong with John,' I said to Mim. 'I cannot understand why he forgot to tell me the case was on and that I was required in court. He was brilliant once we got into it. In fact, I learned a lot about his technique: clear, unrushed, repetitive, carefully outlined, different from his approach with the kids which was aggressive and attacking. But I think there is something wrong; not sure what. This time he was slow and structured. Initially I thought he was not prepared, but he was. But something was off with him. He was distant, vacant.'

A few days later when I went out to John's office to pay his bill and share in the mutual pride of our win, he suddenly phoned the lawyer from the other side to chat, really to gloat, but in a friendly manner. It was interesting for me to see the collegial relationship between the two lawyers despite their having been on opposite sides. On speaker phone the other side was not perturbed. He congratulated John. He laughed and said, 'You win some you lose some.' I knew from John that this lawyer had never lost any before and I was listening for false humility in his voice, but there was none. This had been a very important career win for John against the particular area of law's finest advocate. He was chuffed. On such a high.

Because of this competent performance in the court, his ruthless cross-examination of the witness, and the fact that we won, I did not think too much more about it, until, that is, when I was overseas later and heard that John had been charged with embezzlement and was about to be on the wrong side of the courtroom for the first time, as a criminal defendant. He had 'dipped into' his law firm's trust fund. A lawyer partner who John brought into his practice to help his struggling business, someone John supported and did something good for, dobbed him in. It was now my turn to try to help him.

Knowing John as I did, I ruminated on why he might illegally access his trust fund. It was so out of character. That is when I remembered my feelings about John when we were last in court and the comments I made to Mim that he was not himself in some difficult-to-define way.

After my car accident on the Little Den Hill, I was laid up for about a year recovering from depression, triggered by the concussion. At the onset of that I did something bizarre. I opened my cheque book and just wrote dozens of cheques to pay accounts, far more than the funds in the bank. It made complete sense to me at the time and there is no way I can understand what I must have been thinking. My brain chemistry was not functioning correctly, a symptom of the depression. I immediately knew what was going on with John and how to save him.

From overseas I contacted John and asked if I could help in his defence with strategy. The offer was rejected. I made it clear to John that in my mind, he was suffering from undiagnosed clinical depression. I described to him my car accident and my own experience dealing with depression. I said to John that I now recognised the symptoms in his behaviour during the Fair Work Tribunal case. I knew then something was wrong, and now I knew what it was. I felt it was his sudden skyrocketing to such dizzying heights from the Port Arthur incident that had affected him clinically. I suggested that this medical condition might be a worthwhile defence, to explain his actions. I would even witness his behaviour in my experience. It is well known that poor choices and decisions are made by clinically depressed people. Decisions that appear rational at the time to the sufferer. My ideas were rejected. John is an alpha male and an alpha male has difficulty admitting to weaknesses. John lost his court case and was gaoled, debarred and lost his law practice.

I visited John in gaol and I have visited John and his family since, stayed as a guest at his house, and discussed all of this. Much later, John admitted that perhaps it might have been a good strategy. He was prosecuted by then state Director of Public Prosecutions, Tim Ellis. In my assessment, Ellis is like a pugnacious, aggressive, ruthless attacker, with no concern except for a win, who may have convicted innocent people. He has killed someone in a road accident by crossing to the wrong side of the road. Heavy-set, barrel-chested, a poor loser, he has unsuccessfully challenged many of his courtroom losses. He even appealed the light sentence for his conviction for negligence that killed innocent, beautiful, 27-year-old Natalie Pern in 2013. I guess there are lawyers and there are lawyers.

CHAPTER 28

MELTING POT

If I had been asked in 2012 how to fix a company, I'd have rattled off a confident formula. By then I'd completely transitioned from my technology-driven world of robotics to my commercially-centric world of company turnarounds and had perfected my method, or so I thought. It shows how wrong you can be.

My knowledge of how to fix companies came quickly, a rapid learning curve that sometimes I forget I went through. I recall vividly each of a handful of lessons leading to the set of rules I use to fix companies today. Rules such as: companies are machines for selling stuff, composed of only a handful of parts, and so failed companies are broken machines with some of those parts needing repair, which indeed is how I see it still. Those lessons, more like insights, came surreptitiously in my first few turnaround challenges, so it was certainly a rapid course, in both senses of 'course'.

There is always an apprenticeship preceding a professional endeavour. My apprenticeship was not recognised for what it was until afterwards, so it was not identifiable as a training ground at the time, just an organic progression, an accumulation of commercial and corporate experiences in parallel with my technology ones.

The first lessons came with the first company I worked in after quitting robotics, an accounting firm of all things. A large private one in Sydney called True North Group (TNG), run by the accountant son of its accountant founder. It is where I discovered two rules. The first was about corporate structure; the second, which was closely connected and came as a result of the first, was more deceptive, more subtle, more nuanced. Took more to put into perspective. It became apparent that good accountants do not necessarily make good company leaders. In fact, it came as a shock. In time I discovered that this rule applied generally to many professions. Lawyers, architects, consultants, scientists, inventors, all rarely make good executive managers; in other words, they make poor CEOs.

Many expressions like Murphy's Law, Parkinson's Law, and The Peter Principle were bandied about wantonly at meetings going back to my days in state and federal government. As a novice, they were interesting rules of thumb that could have been urban myths for all I knew, 'old wives' tales', or expressions with origins lost in time. Being who I am, a back-to-basics person, I always delved into them, obtaining and reading the original texts, understanding intimately their meanings and more importantly their application. It was interesting to me to learn that few of the people espousing those sayings had any deep insight to them. No one I knew knew who Laurence Peter was, for instance, let alone had read his book.

Armed with this acute awareness of The Peter Principle – 'that individuals in a company are promoted to their level of incompetency' – it behoved me to go further and learn why. Dr Peter had shown that it was a symptom of promotion on merit from previous positions until their lack of further competency prevented further promotion, therefore leaving that individual in a job they cannot do. But that was not enough to satisfy my insatiable curiosity. It supposed that individuals had an innate limit, like a ceiling above which they would no longer be competent. And more, that training or further training would not solve the problem, but because of their successful track record up till then, they were not demoted or fired. Dr Peter admonished that we come across these senior managers everywhere with consequences. The opening chapter of his book talks about the incompetent design of rooms in a hotel where sounds flow through walls. He imagines an architect somewhere in charge of approving the design, unable to get it right.

This was still not enough for me. Just as I needed to know every conceivable aspect of the world of robotics at the start of my technology career, I was interested in how The Peter Principle related to what I discovered at TNG, whether individuals indeed had a limit to their capacity, even smart professionals. I know there are things I cannot do, because I have not yet done them or not been trained to do them, but I feel that I could. There are other things I cannot do, could never be able to do, because I know they are beyond my brain's capacity, or to a lesser extent beyond my interest level, so I would never embark on those endeavours, like I could never be a maths genius despite being good at maths. What I was discovering was epiphanic, revolutionary. It was the trigger to my complete confidence about company turnarounds, leading to my statement that any company can be turned around. It frustrates me when I see good businesses go under, as the fix is so obvious to me.

My epiphany came from the merging of The Peter Principle with what I was uncovering at the accounting practice, struggling under the leadership of the clearly very accomplished accountant founders.

A digression to the old career advisors one meets on graduating from school, the vocational experts who tell you what you are suited for from aptitude tests. If you are good at details, dotting the 'i's' or crossing the 't's' then you will make a good accountant. If you are a dreamer, an idealist, a visionary, then you will be a good scientist. Those who were neither of those things were the unskilled labourers of tomorrow. There wasn't any career advice for those who were both. My epiphany was that both Dr Peter and the aptitude tests were right.

I discovered, to my surprise, that all good companies have two types of managers: accountant or general manager types who are good at details and persevere endlessly over pages and pages of figures or reports or tactics, and a visionary one, like the CEO who plots and plans and implements a vision and strategy for the company. Having both types of managers is crucial to a good organisation, but they are different species. A good top executive rarely makes a good general manager. Jack Welch from GE could never have run a tiny startup; he probably did not know how many sugars he had in his coffee. And a good general manager rarely makes a good CEO, as the career advisor predicts and Dr Peter described. A professional, like an accountant or a scientist, is the meticulous specialist checking the i's and the t's, caught up in the trivia. A great inventor is often replaced by a different CEO by the investors as their invention transitions to a real company, which needs management requiring different skills.

I think I am one of those trapped with both sets of skills. I see visions, create strategies and plans to achieve them, but then am just as comfortable settling down to the monotonous work of tedious implementation.

At TNG, the accountant founder, an extremely competent financial guru, was nevertheless a general manager who had promoted himself to CEO, where he struggled. The board brought me in. That is how it goes. The talent and skill of governance resides in the board, and when the business fails to prosper, their job is to make the necessary adjustments, including changing the CEO if that is the problem. When private equity or IPO investors participate in a company, they usually place a representative on the board to protect their interests, and that is why many entrepreneurial startups move the inventor out or sideways as the business grows.

This exploration of the distinction between general and executive management led me into the realm of corporate structure, something I previously knew only in a restricted, vague sense, nothing theoretical or analytical, and even that was limited to my own robotics company. I was keen now to understand more, and typically, I read every book on corporate structure I could find. They emphasised that form follows function, structure follows purpose. Reference books and articles presented dozens of examples of the apparently different organisational structures of their example companies. They certainly looked different from each other. Each company, according to this accepted wisdom, had a different structure, was different according to each industry. The closer I examined the structures, though, and aligned them with my own experiences, the more I had to disagree. By now I had also participated in the turnaround of a couple of TNG's clients, very different organisations from each other and from TNG itself and from my robotics company. I disagreed so fervently that it constituted the core of what I did for the next 20 years.

My evaluation, actually re-evaluation, of organisation charts led me to my second insight. They were really all the same. Yes, there were different names for company divisions, different emphases on importance, but the same. Sure, operations might be called manufacturing in one company and service delivery in a non-manufacturing company, but in all cases, it was the division with the function of producing and delivering the product or service. And sales might be called business development or sales and marketing, but it was the same function to get revenues from customers. And marketing might be called communications, but both are the company's method of identifying who the customer was. And size was important. Manufacturing is everything in, say, an automotive company, and next to nothing in an insurance company. R&D is everything in a pharmaceutical company and nothing in a civil engineering company. Focussing too much on the details of an organisational chart, resulted in complex networks of jobs and positions that disguised that at the top there really was only a small number of consistent organisational functions.

Before accepting my own analysis, where did I go? Back to basics, of course. Surprisingly, the basics took me out of my house back to the House of Medici in medieval Florence. The birthplace of companies as we think of them today. For a hundred years from the 1400s, initially through the formation of a bank, ultimately to become the wealthiest family in Europe, the Medici family reign coincided with events involving two famous individuals. One was a fellow Tuscan, Machiavelli; the other was a Portuguese named Magellan. The three Ms.

Banks and lenders existed before, on a small scale because real wealth resided in the royal families. Anything of serious funding was done, or not done, with the approval and participation of the nobility. Magellan's dispute with King Manuel I of Portugal, who refused to support his trip to the Spice Islands, is a prime example. Instead King Charles I of Spain's funding was the reason he travelled west around South America instead of east around Africa. Otherwise he might not have circumnavigated the globe, or at least his expedition without him might not have in 1522.

With the increasing technology of navigation and ships, and the discovery of parts of the world and its riches previously unknown to Europe, exploration indistinguishable from trading accelerated. That's where Magellan, and his ilk, fits in. Funding these risky long-term ventures was now beyond the capacity of even royal families, so the scene was set for the clever nurturing of a private bank. Doing so, though, necessitated the relationship skills to navigate not the sea but the Papal See, and the other parts of the powerful establishment. That's where Machiavelli and his ilk fit in.

Machiavelli is famous, or infamous, today through the use of his name to describe deception through manipulation. His ideas on lack of loyalty are seen in the way he swapped sides unemotionally according to his self-interest. His ideas on ambition and the competition are seen in the way the Medici intermarried according to power grabbing desires. Of these and other psychological devices of Machiavellianism the 'divide and conquer' one is paramount. Keeping control of information, turning sides against each other, creating factions all to ensure centralisation of power.

That is what is happening in the world as I write. Divide and conquer. Nothing is more stable, more powerful than large groups. So, a strategy to weaken your opponent if they are a large group is to split them up. A stable, solid bloc like the European Union or the United States, or ASEAN, is best defeated by making them disintegrate from the inside, to make parts like England break away, or Texas to break away. It seems to be a human trait that people cannot get on with other people and a Machiavellian technique is to manipulate people's minds so that they think they decided to break away.

Despite Machiavelli's diabolical ideas, not all division is detrimental. His advice that it be applied to the opposition seems sensible. Outside a modern company it leads to the antitrust breakup of companies to prevent monopolies. Inside a modern company, though, it can appear in the form of separation of duties, which will in turn bring us back to organisational structure.

Machiavelli reappears in a more nefarious way in modern organisations. The Corporations Acts of most countries are similar. They require the leadership of a company to preserve and maximise the wealth of the owners, the governance and management structures are to protect the shareholders' investments. To not do so is a dereliction of duty. This has been interpreted in many organisations, particularly large multinationals where the shareholders are themselves corporations, to imply growth and profit by any means possible. So, individuals with high Machiavellian tendencies tend towards corporate leadership. Understandably, this comes with all the symptoms of deceit. Gideon Haigh, who normally writes about cricket, describes his moral outrage at this phenomenon so vividly in the June 2003 *Quarterly Essay* article 'Bad Company, the Cult of the CEO'. He describes the advent of the obscene salaries for top executives that have emerged since World War II. I've provided copies of his essay to many of my co-directors.

The existence of the Medici Bank coincided with the need to fund huge projects. Mostly commercial ones, fitting ships with crews and cargo, waiting for them to spend a year or more at sea, their return not guaranteed and even then, with no surety of a profit for the investment. The banks funds were deposits and investments, sometimes even from royal families, and sharing of risk was the key. How to share the risk of such ventures? Entities like this to share risk had been around in various forms since antiquity, but the emerging concept of a company took this idea one step further. It introduced the concept of limited risk through limiting the liability of each participant, and they, in turn, accepting that limited risk. That had not existed before and the idea needed a solution.

The method for achieving this turns out to be world-shattering but surprisingly simple. It treats the company as if it were a person. When you incorporate a company it is like giving birth to a new person, creating a new entity. The risk, and benefit, of the activities of the company, the risk of a ship sinking, or lost freight, or mutiny, or returned goods not selling, are all with the new individual, the company, not with the shareholders who have already satisfied their risk level in the investment already made, their equity. If losses or liabilities are greater, the shareholder still has no further exposure. Establishment of the company is through individuals injecting something, usually cash, according to their desire, and that being the extent of their risk. That is all that they can lose. If you invest, say, a thousand dollars in a company and it loses millions, you still only lose your thousand and are not liable for the quantum of the losses. The same apportionment then applies to any profit generated from the enterprise. This concept of Limited Liability is one of the great inventions of human thought, up there with democracy and justice, concepts that do not exist in nature. A company looks, feels and lives like a person. It is born, has a name and a place of residence, a purpose in life, it can own and sell stuff, must obey the laws, can sue and be sued, has to pay taxes, can acquire debt, just like you and me.

A company has three distinctions from a real person, though. It does not necessarily have to die. Of course, most do eventually close down, but they don't have to; they are designed to live in perpetuity. And the second thing is that, of course, they are not actually a real person, so they cannot make decisions themselves but must have a system to make decisions for them – their governance. Like a person in a coma has a power of attorney. This is so critical that it is written into all company laws in all countries. It is their board of directors. And, thirdly, companies cannot vote. Because a company is actually a group of people, who already have a vote, a vote by the company would give them another vote, which is illegal. Companies should therefore be politically neutral.

I am against companies taking a political position or influencing the politics of their staff.

All of this is critical to my main point. I was talking about organisational structure. Companies with this amazing concept of limited liability were made so that hugely expensive trading expeditions could be funded by a number of people contributing various proportions and sharing the risk and the benefits in those same proportions. It was the opportunity for trade that stimulated the creation of these ideas and turned the concepts into something new, a structure to buy and sell.

A company is a machine for selling stuff. That is exactly what it is and that is all it is. The optimal machine for doing this has been honed over the interim 500 years, to a well-defined structure. It is just like an automobile; after all the trial and error in the beginning of the invention of the automobile they all now have essentially the same structure, a perfect structure for carrying you and your cargo from A to B. They have an engine for power; a transmission to get the power to the wheels; a control system of accelerator, brakes, indicators internal and external; a chassis; a fuel structure; and a compartment for you and your goods.

It is similar with companies. They are machines. A company, all companies, have seven components, a couple more than the handful I alluded to earlier. A company consists of R&D, Marketing, Sales, Operations, Administration, Executive Management, Governance. That structure, when working, is like a well-oiled machine. When one part is broken, the machine malfunctions, but the fix is then easy, as it is specific to the function of that piece. When a company is broken, needs a turnaround, it is not the company that is broken, it is a component of the machine that needs repair. Consultants and executives going in wholesale making radical changes often use a blind shotgun approach, wasting time and money.

This set of seven parts is what I call Standard Structure. It is where I first espoused that if you have Standard Structure, it does not guarantee that you will be a successful company, but if you don't have Standard Structure, it guarantees that you will fail.

This melting pot of ideas and others was the result of all my research and analysis of what was wrong with the accounting firm, my first job after robotics and my first turnaround. Or at least that's what I thought then.

One of my favourite little management books is *The E Myth Revisited* by Michael E. Gerber. It is only a primer to my formula. In my management workshops I sometimes walk through a similar scenario – a fictious chocolate chip cookie startup. The founder makes cookies better than anyone her family knows, so they say she should start a business. Her R&D is done in the development of her fabulous recipe. But it is no use having the world's best chocolate chip cookies unless the world knows about it, so market it, tell the world. If marketing is successful, there will be requests for cookies, so sell it by taking orders and taking payments. That gives an indication of demand, preparing for production. Now make them, the operations of the new business. All of this needs administering. By law there needs to be someone to take responsibility, the board of directors, to govern the activities of the company, pay wages, pay taxes, obey laws, establish a vision and plan for the successful continuation of business.

The roles of management and governance are different. The board determines the company's vision according to shareholder wishes and management makes it happen. Management is accountable to the board. The conduit between the company and its board is the appointed chief executive officer, delegated the board's hands-on management duties, separating governance and management so that there is a true independent oversight. Companies in which the CEO is also the Chair of the Board are not independently oversighted.

Sometimes I add an eighth component, the community. The community of people from which the investors come, the employees come, the product or services go to, and who are affected by the activities of the company. That feedback loop of people from the community coming into the company and the company in turn affecting the community should make companies respect their communities and become good corporate citizens.

The benefit of this structure becomes evident when the company struggles. The problem is going to be in one or more of the components. Diagnostics identify where, and repair tools fix it. The number of components is finite, so diagnostics are quick and repairs are easy because of the partitioned nature of that component – sales, marketing or whatever. If a car has steering problems, the mechanic does not waste time checking the turn indicators.

Simplistically there are fundamental rules of thumb for each part. Marketing should always be done independently, sales is a numbers game, operations is a cost accounting exercise, R&D should have the next product ready to go at any time, people really are your best asset.

That last is the most important rule and it surfaced quickly in my first few turnarounds. I discovered that most companies in strife were short of people by the time I was brought in. Those with good resumes or prospects had often already departed. Others could not leave the job or work in another region because of personal circumstances. So, the remaining people were perhaps not the most talented, or most motivated. But no one turns around a company in isolation – people are needed. A key need became what I term a 'talent audit': to assess who I have, where to locate them, whether they needed to be relocated or retrained. At times people have been in a job not matching their skills, and retraining people is key. Half of that task was earning trust, which was done through walking the talk and generating credibility. No one knows what the other half is; that amorphous thing called 'leadership'. I think 'leadership' is the ability to generate followers.

I resolutely believe that solving financial or commercial stress in a company is never solved by sacking people. It was to become my cornerstone principle, my trademark approach. Every company manager I know says that people are their greatest asset, yet when profits are down and shareholders scream, it is the people they sack. The salary bill is usually the largest expense of a company, so it is a quick fix, nothing ingenious about it; a bookkeeper can achieve as much. Sure the net financials at the end of that year might show an improvement, on paper. It is often the way of short-term consultants and bean counters who come in to supposedly help a company, but the short fix is unsustainable, because the next year there is no one to sell the product, no one to make the stuff. The company leaders should instead be questioning themselves as to what it is about them, about their management, that has led them to decide that their people, their espoused best assets, are the first assets to let go. They should be blaming themselves and not their people. Maybe even sack themselves.

At the time of writing, Elon Musk has taken the helm of Twitter and sacked half of their 7500 workforce. He seems to ignore the obvious: these are the people who built the company. They are the company. That's what the word company means. In 1996, Al 'Hatchet' Dunlap was hired to turn around Sunbeam. Sunbeam was a prized American company, a hundred years old when Dunlap got hold of it. He immediately sacked half of the 12,000 staff. Of course, with such a reduced salary bill, the expenses were down that year and the company showed a profit. After that, though, there was no one to make stuff, no one to sell stuff, no one to check stuff, like quality or finances, and no one to service customers. So, it was a short-term fix that looked good for five minutes. This approach destroyed Sunbeam. Similar to what has been happening at Twitter. Such drastic cuts earned Dunlap the title of Chainsaw Al. But that was nothing: it turned out that all his financials were fabricated. He fabricated production figures, sales, expenses, anything to make his performance look good. He was sued by the SEC and shareholders and ended up in obscurity. Amazingly he had done exactly the same thing in the two companies he managed before Sunbeam, both known ahead of his hiring. The true problem, in my opinion, is the incompetent board of directors who hired such a ruthless, arrogant, abusive manager. That is what Gideon Haigh says. Anyone can do that. A more competent manager, a true leader, a clever person worth their CEO salary, would be skilled at ensuring those people are trained, led and nurtured to be the great assets they are purported to be. Leading more people to make more profits, not sacking people to reduce expenses.

There is one final piece of the puzzle, a fundamental part of what I discovered to be crucial to dealing with stressed companies. It was the company life cycle research of Larry E. Greiner in the *Harvard Business Review* of July – August 1973. It showed that companies progress through various inevitable stages as they mature and grow, at any stage of which they could, and most do, fail. Those that survive a stage have demonstrable internal changes that he describes. The work was different from the previously accepted model of 'growth-maturity-decline.' I was able to integrate his theory into my Standard Structure theory.

That was my melting pot of discoveries, ideas, contradictions to accepted wisdom, back-to-basics research; formular for turning around companies.

There is a great song by Blue Mink called The Melting Pot. It tries to describe a formular for assimilating people, of living together in harmony. Even at the time of release in 1969 it had complaints for racial undertones, but the thought was good and the solution seemed good too: to blend everyone over time till the human race was 'coffee coloured.' But I liked the melody and the sentiment.

Today, from my global experiences, I think differently. I don't want assimilation, I want diversity, variation, choice, difference.

So, in 2012, if I'd been asked how to fix a company, that is what I'd have described, a melting pot of ideas into one neat mixture. It is what I had done successfully for dozens of companies by then and dozens more divisions of companies. But just when you think you know it all, something comes along to knock the wind out of your sails.

CHAPTER 29

AD HOMINEM

The something that knocked the wind out of my sails was working with a company called Krucible Metals Ltd, curiously also a melting pot appellation, a junior explorer in the tropical north of Australia searching for phosphate and rare earth metals. Both crucial minerals, one for fertiliser, the other used in every known advanced technology.

Krucible was listed on the Australian Stock Exchange, had raised a lot of money in its successful IPO, was created by a competent geologist and had discovered phosphates and rare earths, but its share price had been in decline for many years. That was because it made the discoveries at the height of the Global Financial Crisis and the worst mining downturn in recent memory, a double whammy that prevented the good prospecting from generating share price increases and more fundraising. I had just turned around a mining company in Tasmania and was headhunted by the board to turn Krucible around.

My proven melting pot of turnaround ideas was brought to the fore.

Here was a professional, a geologist, excellent at his job, more than capable of creating and establishing his startup company, but when things got tougher it was beyond his skill to act as an executive manager, the one required to plan strategies and vision for sustaining the company. He and the board were appointing their first professional external CEO.

People are my first priority and almost universally the first thing I do is my talent audit. That was complicated by the sudden and unexpected departure of the chairman and two directors from the board as soon as I was in place. So, before reviewing my staff there was a need to attract new directors, including the chairperson. In my view, companies who have the CEO also as their chair are poorly governed, so I refused to take the chair position myself. A good board consists of a mix of talents: finance, legal, marketing, business, industry knowledge and so on. We were lucky to attract a lawyer and an industry spokesperson to supplement the remaining directors.

It turned out that Krucible had been searching for a new senior geologist, overlooking the junior geologist who had actually made the mineral discoveries. I promoted her to the top position and made other staff permanent employees, to garnish trust and loyalty and credibility. Despite the poor financial position, staff were given overdue pay rises, requiring directors including myself as managing director to sometimes delay paying ourselves.

My second priority was the finances, which disclosed the reason behind the director departures: the company was almost insolvent. Despite the limited liability associated with companies and the fact that they are obligated to pay their debts, one of the few ways that the people inside a company board can be liable for its debts is if they allow the company to do business while it is insolvent. That meant that to stay legal I had to prepare a contingency close-down plan and budget ensuring all obligations like taxes, staff redundancies and creditors would be covered if we failed to secure additional funding or revenues and ceased business. The plan indicated that we had only two or three months of space before the organisation would have to be wound up.

Taylorism dictates that change is good. I have been lucky enough that share prices have doubled, even increased as much as tenfold just on my joining a company, and there was a small improvement at Krucible, but it was just a blip and short lived. Such was the dire world economic situation and, I was to discover, what I sensed was antagonistic action from some quarters wanting the share price to remain low, not for short selling but other vested reasons. When a broker takes a liking to a company, they are considered a Market Maker, but there are also Market Breakers, almost always anonymous. I had definitely stopped the share price decline, especially compared with industry averages and indices, but it was still not rising no matter what I did.

That meant, clearly, that the next priority was to get some cash quickly. Very quickly. Most listed companies have an advantage over private companies because they are able to go to their existing shareholders and request a top-up of funds. It's called a rights issue, so we did that. Shareholders generally participate because it means it will save their existing investment. Normally this is an expensive exercise, with brokers, investment banks, lawyers, specialists, and a large proportion of any raised funds disappearing to those parties. I had no resources to do any of that. Instead, I trained my key geologist to do a road show with me, in the same way I was trained years ago by Deutsch Bank. The participation was small, but it gave us a much-needed injection of funds. Another few months of life.

The next thing was to stimulate interest in the company. Its plight was very visible, creating negative waves. The way a company makes a splash is usually through media releases, but I wanted to make a tsunami with a lot of noise. My approach was through the conference circuit, and the available funds were put to paying for a schedule of local and international conferences and trade shows. To keep travel and accommodation costs down, I stayed with friends around the world, travelled economy on planes everywhere. In Brisbane we searched for a cheap corner pub hotel room instead of a premier room downtown. That was no sacrifice, that was how I had always spent shareholder funds. It is how I spend my own funds.

In social media, content is king as they say. The same in conference presentations. I had to find a story to convince the market to invest in us, despite the sad economic climate. If that happened, the share price decline would stop and hopefully start to climb again. The company had several things going for it. By now we had a new energetic board, with a full gamut of experience: legal, industry, technical, business. We had a new CEO, me. And we had the recent discoveries of premier minerals. Phosphate cannot be replaced by any other fertiliser and without rare earth metals, no mobile phones, coloured lights, tough alloys, super magnets or miniature electric motors, but so far our discoveries had not generated much interest. So, the conference presentation material was brought up to date. Instead of a technical presentation like every other mining company on the planet does, I presented a case for why investors should invest in Krucible instead of any other explorer. This was market differentiation. If you have the best chocolate chip cookies in the world, tell the world. But tell them how you are the best, what is different.

The first local conference resulted in a call from a Singapore/ Indian company that had been approached unsuccessfully before by the previous management, offering to form a joint venture with us to mine the phosphate discoveries. This was a breakthrough. I signed that deal ravenously. All such contracts have a due diligence clause – DD – which means there is a cooling-off period while each side checks and double checks the accuracy of disclosures and facts. Krucible struggled to get our company through the third party's due diligence processes. As soon as time was near to inject some money, the other company would call on a part of the contract that allowed them to extend the DD time. There was nothing wrong with our disclosures, so it was confusing as to why they kept delaying. One thought was that they in fact may not have had the funds themselves. But I had few options, so we kept extending the JV contract. It did not take long to determine that the company was actually trying to break us, knowing we were desperate for cash, expecting us to eventually accept a fire sale alternative offer. My conference presentations were starting to generate additional interest and new opportunities were appearing, but the JV contract was exclusive and we were prevented from entering into any other agreement. We were locked in.

The next call was from a New York private equity fund that had also seen my conference presentations. They were prepared to inject a couple million into us, in order to bridge us to the JV completion. As an equity investment, it was not a commercial deal like the JV. They met us in the state capital city of Brisbane and said they invested in new management and were hands-off. That suited me; I was the hands-on guy in our company. The $2 million could be drawn down piecemeal. In the end I did not need it all, resulting eventually in only a 20 per cent dilution of shareholder value. A small price to pay to save all past investments.

These funds gave us breathing space and before I embarked on the international part of the conference circuit I received the third call. This was a Chinese company also interested in the phosphate discoveries, but they wanted to buy the tenements together with the phosphate rights and the work in progress, which included the mine licence and the associated clearance of native title, cultural heritage, and land owner rights that my directors had negotiated. Normally I am against asset sales, which I also see as short-term fixes, but it too conflicted with the JV contract. The trouble was that I was still locked into the earlier JV contract which was still threatening to break me as they dragged on and on, costing us time and money.

During all of this, we restarted our core business – more exploration and discoveries. Under the circumstances, the share price stabilisation was still a positive result that had started to be visible to the market and reported in financial newspapers.

Now that I had an alternative asset sale cash option, I pored over the earlier JV contract to find a way to get out of it. The contract was still being dragged on by the other company. Suddenly I saw the escape route. It was the same ploy they had been using against me. I was about to give them some of their own medicine. With the second contract ready to sign, I waited for the company to tell me that they wanted to extend the DD yet again. They were flabbergasted when I said no and refused an extension, which effectively ended the agreement. We quickly entered into the asset sale agreement with the Chinese company for about $13 million. We had sold a third of our assets for double the value of our market capital. We were home. Or so I thought.

In turn, the new asset sale contract led to additional benefits. The purchaser was only interested in the phosphate. It left all the non-phosphate mineral rights to us including the valuable rare earth metals. On top of that, they were interested in hiring us to do the mining for them. It was almost like selling it and keeping it. It was a bit like selling it and keeping it.

I finally embarked on the international conference circuit, but before doing so, I took care of one final matter which led to a further refinement of our corporate conference material. Our interest now was primarily rare earth minerals. These metals are a group of elements high on the periodic table, with similar heavy atomic weights, making them extremely difficult and expensive to separate. They are not rare as a group, but rare to separate individually. Once separated they have amazing properties and are used in every aspect of the most advanced technologies today, everything from mobile phones, to colour TVs, to automotive, to wind turbines, to rockets, to jet engines, to defence missiles. The world would be different without them. The world's rare earth industry is dominated by a Chinese monopoly and outside of China there were about three rare earth companies, each struggling. Back home, Ray, one of our directors, had initiated an inexpensive, novel rare earths preprocessing technology at ambient temperatures and pressures. On my international conference circuit I visited processing companies in Mountain Pass in California, Cheng Di in China and Rochelle in France, to compare what we had with the expensive, polluting existing technologies. Once we were mining and processing, we would easily compete with the Chinese.

With millions of dollars in cash-on-hand we invested in our own freehold premises instead of renting, and as a long term investment, started looking to buy profitable resource companies, and negotiated long term investment deposits for the extra cash.

With marketing being taken care of through conferences and printed media and social media content, my attention turned to sales. I was working through the parts of the machine.

With high worldwide demand and limited supply, the market price of rare earth metals should be sky high, but they were not, never had been. Pundits constantly predicted their time will come and the commodity prices will start to rise. My research showed that that had never happened and never would. Being who I am, back to basics, I had to understand more, especially if I was going to be developing these assets for Krucible. On the surface it made sense that the prices should be higher. The result was a new economic paper called 'The Global Dynamics of Rare Earth Metals', which I wrote and published. It showed that unlike most advanced technology commodities that are used in high value sectors, rare earths are mostly used in a preponderance of consumer products, which by definition have a modest price ceiling, beyond which alternative manufacturing and component options dominate. They are in what is called an elastic market. The work was published in a peer reviewed academic journal, adopted by industry leaders like the USA critical Materials Institute, and written up in financial newspapers and trade journals. This became the new theme for the conference presentations and finally saw our share price start to recover. The turnaround had taken almost three years, was perhaps my most difficult but satisfying turnaround project, but was when everything went wrong.

With cash in hand, share price finally starting to rise, our continuing mineral discoveries, an unprecedented understanding of rare earth pricing dynamics, proprietary and competitive processing technology, brand new freehold premises and M&A plans in the works, we were poised to be the premier rare earths company outside of China.

But Australian corporate law has an anomaly. There is a section that allows any shareholder, for any reason, to call a special meeting of shareholders, an EGM. It is used most often to remove directors. Directors cannot be sacked. Otherwise, boards could easily be stacked, in the interests of a particular shareholder, for instance. Normally a director retires, or has a limited term according to the company constitution, or is not voted back in at an AGM. The s240D section of the Australian *Corporations Act* allows a way to remove a director. It is an anomaly because it disrupts the normal course of events. It might occur in the middle of a strategy that has not come to fruition yet, and it gives too much power to selected shareholders because anyone with as little as 5 per cent shareholding can call a shareholders' meeting to vote a director out.

With our turnaround complete, it was inconceivable to me that anyone would be disgruntled, but there it is, there is the missing part of my melting pot formula. The abuse of ad hominem. The part not catered to is rationality. A formula is a logical thing dependent on rationality, but unfortunately people can be irrational. It is the ability of individuals to bite off their nose to spite their face. But usually they have to be misled, lied to, manipulated. Ad hominem means to attack the person rather than their actions, which is what happened, saying my background of achievement was all a lie, such that some shareholders became suspicious and voted against me.

I believe the shareholder who called the action wanted to control the company because of all of its cash. He had approached the board requesting a board position, which was rejected because of his mediocre history and lack of adding anything relevant to the board. Perhaps to be vindictive, he then attacked the board, threatened at least one director and intimidated others so that they resigned. I was the only remaining director. To get enough other shareholders to vote with him at the EGM, he offered them a share of the huge cash funds and told them he could do a better job than I had done. The nose and face analogy is a euphemism for short-term gain despite long-term loss. The shareholders voted me out, by just a few votes but that is no consolation.

This had never happened to me before, or since; usually the opposite occurs. Shareholders are usually grateful, ecstatic, relieved. Sometimes there is professional jealousy. Sometimes the ease with which I fix a problem that defeated those before me is hard for them to take, but that was not the case here. Because of the desperation on my appointment, no time to lose, I did not cater to this event in my contract, so I was exposed and unprotected. After I left the company, its share nosedived again and ten years later has never recovered. Those shareholders who voted me out have lost everything. With the one-off cash divvied out, the company has constantly run short of money and shareholders have lost some 90 per cent of their value through continued dilution from fundraising. All of my commercial deals were shelved. The premises were sold for a quick cash injection. In terms of its purpose in life, the company has neither accomplished any mining nor made any new commercially relevant discoveries.

If I am asked how to turn around a company today, I'd spell out my formula and add that sometimes that is not enough to satisfy some people, and add additional protection to cater to shareholder stupidity.

CHAPTER 30

SUPER BIKES

At Princess Street Primary School, in the classy suburb of Sandy Bay, one of my classmates was Peter Risk. He was Dutch, so it may have been Pieter Riske. Another who lived behind us was a Canadian, son of a university lecturer, and one was the son of the Dr Cannon who used to treat my asthma, and one was Wendy Chandler the daughter of a wealthy business family with a large successful garden nursery. Although exotic Europeans or Americans, professionals, these were different people from the refugees and post-war immigrants encountered at Waddamana. These were not Polish or Greek unskilled or semi-skilled workers. The school was adjacent to the newly built university and these were the children of lecturers and professors and doctors and lawyers, business leaders. Actually, Australia's oldest tertiary institution starting off as Christ College in 1846, but by then on its new campus, an amazing outer suburban location with room to expand and breath that reminds me of Stanford in California. I would go with Peter to his house after school to play. He had different toys, and it was the first time I learned that Donald Duck and Mickey Mouse could speak Dutch in his Disney comics. My exposure to different nationalities and races and cultures and languages and especially foods was a good primer for what was to come in my life. In particular I have no recollection of any racism – these people were just interesting friends and neighbours. Names like Djbroski, Banovic and Kesamechek rolled off my tongue as easily as Smith or Brown. As of yet there were no Chans or Kumars or Mahmouds, though. Those were to come later.

Mim's sister Keryn was married to a Dutchman named Piet Blokker, and his school friend Herman Stevens worked for Royal Dutch Airlines, so Herman could fly at discounted prices and he visited his friend in Tasmania often, which is where I met him and his wife-to-be Michelle, also Dutch, but who grew up in Canberra.

On my early trips to Europe I'd visit them in Holland, in the village of Hoofddorp, near the Schiphol Airport where Herman worked. My observations while touring around the towns and cities in Holland germinated my economic ideas on the difference between growth and development. I'm one of the few people I know who do not want more people in Tasmania. I say if you want a large population, anywhere else in the world will suit you fine, just go there. Places with sparse populations are rare.

Places with sparse populations with a perfect climate and perfect habitat are even rarer and should be preserved, like the old growth forests in Tasmania should be preserved. More people means more pollution, more lines, more traffic, more delays, fewer parking spaces, less speedy health care, crowded schools, shortages at the market, more frustration, agitation, crime. In my view, the only ones benefitting from more population are armies for cannon fodder, religions for converts and big business for profits. We learned about natural population control at university, how disease, starvation, catastrophe and conflict work to stabilise populations of species, keep them in equilibrium. That balancing act is visibly happening with human populations, with hundreds of thousands dying in a tsunami, hundreds of thousands dying through drought and flood and heatwaves, hundreds of thousands dying because of food shortages, hundreds of thousands dying through pandemics and plagues, millions dying through wars and battles. Some countries with the larger populations finally have birthrates levelling off and yet there are still cries to increase the numbers. I don't get it.

Holland is the most densely populated country on earth. Instead of promoting more and more population, a growth strategy, they constantly improve the infrastructure and standard of living for the people already there, a population that is more or less fixed. So, the country develops its infrastructure. A simple example I saw was the traffic lights at intersections, where a red or green light well before the traffic intersection indicates if the lights will be in your favour or not by the time you get there, so you can continue without slowing down, making traffic flow much more efficient and therefore less polluting, and therefore the population more healthy, so less need for hospitals and aged care facilities. Another was the electric buses with induction transformers under stopping places on the road, so the batteries trickle charged constantly while on route. Or the emphasis on using high intensity agricultural technologies, hydroponics for instance, instead of clearing forests for more acreage. The Dutch have one of the oldest life expectancy populations in the world – around 83. Only idiots offer incentives to have more and more babies in my view.

I always heard Herman Stevens's name pronounced the English way, and it was fascinating to hear Michelle answer her home phone one day and pronounce it as 'Hairman Stayvans'.

Herman had a side line designing, building and selling reclining bikes, super bikes I call them, those bikes where the rider pedals them with hands and feet from a reclining position, either facing down or up, prone or supine. Not since my high school days had I ever been a cyclist, but the design and technology of his bikes fascinated me. He was an avid cyclist, like all Dutch.

On another trip I stopped at their house for a couple of days because there was an international trade show called Interclean to be held that year in Amsterdam, and I planned to visit it. The annual tulip festival was also on, so a day trip to the most spectacular fields of colour I have ever seen was planned, with perfect glossy tulips as if mass produced by a production line. Rainbow of colours to make a gay parade envious. Fields as flat as a Dutch pancake, with the immaculate windmills here and there dotting the landscape like conning towers checking everything out. But no similar intensity of perfume; if anything, a sweet, delicate citrusy odour.

Then the next day I went to the trade show. To my amazement I witnessed a large robotic industrial floor cleaner named RoboScrub sweeping around an enclosed floor space autonomously. It was manufactured by Windsor Industries in Denver, Colorado, and had been developed by Denning Mobile Robotics, in Boston. I knew Denning very well. The company was one of my competitors, and some of the people in its founding, including Hans Moravec, who had invited me to CMU a few years earlier, were known to me very well. Denning was a hugely more visible robotics company than mine, and well known for its successful NASDAQ IPO raising some $20 million dollars in 1982 – a lot of money then, as it is now. But I'd never even heard of RoboScrub. It was a surprise because I keep detailed tabs on my competitors. The robot worked well, was visually impressive, and was garnishing a lot of interest. My thoughts were on the comparison of this machine versus my tiny domestic ones for Moulinex and General Electric. Denning had also been the company calling GE when I was there, offering to do their Florbot project for free and urging them to cancel my contract to build it.

This was the introduction to another of my rules about fixing companies. I was to learn about the trap that technology startups fall into: being enamoured with the complexity of their inventions. Inventors, scientists, engineers and CEOs raving, boasting even, about how complex this device is or that design is. Only technologists are excited about complexity. The more complex, the more parts, the more they love their creations. To the consumer it means more parts to go wrong, more expensive to buy and fix. Business plans of such companies focus on the technology as if that is all there is to it, ignoring the sales, and marketing and competition, because they know nothing about selling, they don't believe in marketing because everyone will instantly want to buy this thing, and they believe they have no competition.

We saw it on the Space Shuttle. The use of two fuel cylinders joined by an 'O' ring instead of one cylinder with no joint, no point of weakness halfway along it, which was equally feasible. I was seeing it in RoboScrub. The demonstrators ranted about the number of sensors, the number of computer boards, the miles of wiring, the thousands of lines of software code, as if they were benefits. I immediately had an idea and could not wait to get back to a phone.

I got through to Windsor Industries and described who I was and arranged, more easily and quickly than I had expected, to visit them in Denver. When I eventually arrived, they said this was their lucky day. There was no poker playing here, they were desperate for help. They had invested heavily in RoboScrub, through contracting the robotics technology with Denning, and Denning had gone out of business! Wow! Windsor was stuck with a product and technology that they could not maintain. It was similar to how I discovered Professor Robin Murphy a short while later stuck with her immobile MRV3 research robot from Denning. Now I was in shock. I kept a close eye on my competitors, or so I thought. Not only had RoboScrub been a surprise at Interclean, I had heard nothing about the demise of Denning. I told Windsor that I would be able to help them with RoboScrub, but first I had an idea about Denning and asked for them to be patient and I'd get back to them.

From Denver I flew to Pittsburgh, staying at Jon Jarvik's house, next door to Hans Moravec. They even shared a driveway. Hans had been a key scientific advisor to Denning through its inception and startup, and although not officially a part of the company, was certainly involved and informed. I asked Hans about it and he told me the story. The failure of Denning was as closely held a secret as possible. It had actually closed about nine months earlier. Hans told me there was no one left at the company and there was only one director from the board still involved, who was in the process of wrapping everything up, and that was Robby Long in New York. So, that was my next stop.

On the plane to New York to meet Robby Long, I wrote a turnaround plan literally on a piece of paper, presenting it to Robby over a coffee near his residence in Chelsea, not far from where I stayed much of the time on West 22nd Street with Gordon and Dabni, with the best examples of brownstones in the city. Robby was the only person left doing anything with Denning, representing a large shareholder, Mike Pisani Jr. Essentially Denning's technology was inferior, its robot products did not work effectively, the products were too expensive and received few sales, and the company had no sales or marketing activities to speak of. Their classic and typically arrogant marketing approach was, 'People will call us.' I guess that is great, if they call.

I was more behind the ballgame than I realised. Robby said the company had actually closed down when its debts became unmanageable. As the sole person taking any interest, he was acting like a lone director on a board of one person. The last two employees called him asking permission to fire each other, only half joking. The office had been locked up by the landlord, who was owed around $300,000 in back rent, and no one was officially closing it down. All services were turned off, bills mounted, and back rent was left unpaid. Ex-staff still holding keys entered the premises to steal things of value, including taking office filing cabinets which had the contents emptied and thrown on the floor. Over the freezing Boston winter with no heating all the water pipes froze, then in the thaw, they burst, causing extensive water damage to all the office files rotting on the floor, destroyed. The landlord took the company to court for his unpaid rent, and as no one attended the court hearing on behalf of Denning, the landlord was awarded all the assets of the company including its patents. In effect he now owned the company, or at least controlled it.

For instance, at one time when I sent out letters on the Denning letterhead, I received a warning from his lawyer to desist, because the logo on the letterhead was part of the company's IP, its intangible assets. But that was later. Back in NY that day, Robby asked if I could recover anything from that mess, and of course I said, 'I can do that.'

My turnaround plan required about $2 million to restart Denning. I had never written a turnaround plan and I did not recognise it as such at the time, but that is what it was. In retrospect it was my first company turnaround. In retrospect I now see many of my previous robotics activities as project turnarounds – helping my customers or contracts to recover from failed or stalled development projects.

Common sense suggested that no one would be interested in investing in Denning, a total failure; however, I decided to try a group of existing shareholders who had lost everything who were known to Robby, so they had nothing to lose if a small additional investment would revive the company and restore the value of their original investments. In a desperate attempt I managed to raise about $47,000. Basically nothing. Regardless of how much money I raised, though, I could do nothing without the landlord on side, so that was the first thing to tackle. It turned out that the landlord had not been idle. He had all the assets and technology of Denning and whatever was left of the equipment and materials in the facilities he had rented to them, so had been in discussions with another university with preliminary plans to restart Denning himself. It was easy to convince him otherwise. With a sad commercial history, technology that had underperformed, no staff, no money, and no prospects, his job to rebuild what had taken 11 years and $20 million would have been all but impossible. I could sense from his fatigued body language that he knew it would be futile. I convinced him it was something I could do, because I had unique alternative technology, was in business, had a good track record, and would lead it myself. In good faith I said I would pay him the $47,000 to cancel the debt and release everything. Not content to be too naïve, I still hired a lawyer and we settled things in court. Later the landlord said he could not believe it. One minute he owned Denning, the next minute he said I 'went for the jugular' and he had lost it to me. That's where I learned about going all out in any legal activities. Today I still use the American method: I go for the jugular. It scares the opposition and sometimes it scares my lawyers.

As soon as the deal with the landlord was established, we hired a semitrailer, loaded all of the things from the premises and shipped everything to Pittsburgh. The same shareholders helped with the heavy lifting. Things were finally happening in this reincarnated company. It was interesting to see the container arrive in Pittsburgh hitched to a different tractor with a different driver. To get a container across country quickly, they do what the old stage coaches used to do. They used to swap out teams of horses at coach houses and continue on with a fresh driver. The container was changed over about halfway and continued on non-stop. I only had time to get back to Pittsburgh myself.

I had rented a vacant space in a cheap neighbourhood; the opposite of what Denning had done. Another classic startup mistake, spending money before they made it, leasing a too large, too expensive, too prestigious space before they had earned the right to. I have a photo take on my first day at the new Denning office that shows a deck chair, trestle table and a phone beside me on the floor; me, the only employee in an otherwise empty 1000 square feet space. Inauspicious new beginnings.

Already, from New York before I had negotiated access to Denning's assets, I had taken care of Robin Murphy's Denning robot for her out in Colorado. The next thing I had to do was go back to Colorado to sort out the Windsor situation. They had complaints that the squeegee action left watermarks on the floor. I knew this would be because the software code lifted the squeegee a fraction too early. This was a cross-country trip by car, no funds for flights, again with my trusty soldering iron and multimeter in my toolkit. There was no access to code, software or systems to modify the operating system of the robot. Instead, it needed a hardware bypass solution. With some ancillary circuitry to modify the squeegee action, and approval from Windsor at this evidence that the creator of RoboScrub was back in action, they were happy again with their robot investment. I now had two satisfied customers, probably a record for Denning.

Back in Amsterdam, the most significant thing I did from during of my trips was to visit the Anne Frank house, now a museum, with long, sad and quiet lines of visitors along the street beside the canal, not knowing what to expect, crying awkwardly by the time they exited.

There was another Dutch project. Xerox was in decline like many of the mature technology companies resistant to the changes coming to their industries caused by the internet and soon the iPhones and then social media. Not many people copy a document anymore. They take a quick photo of it with the mobile phone, or they send a pdf electronically. Xerox was upgrading the necessary automation of production of their photocopiers, a kind of death throe. In Europe their factory was in Venlo, an industrial town in Holland bordering Germany. Xerox hired Denning not long after we had established and grown in the new Pittsburgh factory, to design an autonomous conveying robot, an AGV, to carry the devices through the production line. A mistake they made, as did other companies we contracted to around then, was to over-manage their third party contractor. To hire or outsource a project that they could not do, then interfere to try to manage it. It's a vestigial style of large, powerful companies, browbeating their suppliers to save a few more pennies. This is seen in the vertical supply chain integration with massive food supermarkets, paying growers little and monopolising retail prices for their profits. While the micromanagement of their suppliers worked for growers and for old engineering technologies, robotics and similar disruptive technologies were a specialty. Xerox sent us specifications for the robot, which we matched, but instead acquired a cheaper less flexible device that did not meet their own specifications. By 2000 Xerox required its own turnaround.

On yet another visit to Holland I arrived when Herman and Michelle had planned a day out to a scheduled meet for racing these amazing super bikes. I was keen to go with them for the day out, but only as a spectator. After a lot of pressure, though, like an idiot, I agreed to participate in a race. I had had a bit of a ride previously around the streets outside their house to experience a superbike, but was not convinced it was a good idea to join a race. Like most males, though, I succumbed to their patronising, 'You look great on that bike, Allan, like you were born to it.' Later, at the event, everyone had their kit, their Lycra outfits and helmets. I had nothing, so I was fitted at the last minute with some shorts and a t-shirt and a bike helmet someone loaned us. It had also been a mistake to eat a large meal immediately before setting off. At the start I pedalled out as fast as I could, but was immediately left behind the rest of the racers. I turned at least four left-hand corners, thinking 'Well, that's a full 360 degrees, so I must be near the end line,' yet there were more and more corners. Clearly, I was confused about the course, and as I persevered no one was left in sight ahead of me. I was alone. As I neared what I was sure must be the end of the course, I came around another bend and had to start dodging cyclists coming at me flat out in the other direction. I persisted and limped across the finish line to rounds of applause, and possibly cheers of relief that I had not died of exhaustion. The organisers had tired of waiting for me to finish. Assuming I had gone off to the side or something, they started the next race in the opposite direction. It helped to bolster Holland's reputation against the Aussies. They relished in reminding me that the Dutch discovered Tasmania.

CHAPTER 31

AUSTRALIAN RULES

In 2014 I finally met up with Bernie Appel again, the CEO of Tandy/Radio Shack, and accepted his oft repeated invitation to have lunch with him if ever I was in Fort Worth. It was when I was living in Dallas, Texas, which is about 35 miles east of its sister city of Fort Worth, the 'Cow Town' that holds Billy Bob's honky-tonk, the largest cowboy cabaret in the world, that I had been in Fort Worth from time to time. Dallas-Fort Worth is like opposite faces of a coin: modern, swank, cosmopolitan business-oriented Dallas and dusty, Stetson-decked, alligator boots–treaded Fort Worth. Each trying to be like the other. I'd already met Bernie Appel, in Hong Kong, and in fact that was the coincidence that led to my being in Dallas in the first place.

Over a period of 34 years, brusque, corpulent, energetic Bernie was CEO of Tandy Corporation's Radio Shack, building it up to be to the largest electronics retailer in the world. Bernie joined them in Boston in 1959 when they sold catalogues through three stores. When he retired in 1994 as CEO it had 7000 outlets and was worth $2.3 billion. His Russian immigrant–given name was Beryl, but he was known as Mr Radio Shack.

In Australia Radio Shack gave rise to a competitor called Dick Smith, the man and the eponymous company. Dick was a slight, bespectacled man as opposite to Bernie as it was possible to be. Dick was technical, analytical, opinionated, but equally competent commercially. Bernie was seat-of-the-pants, intuitive, decisive, adventurous. The Aero Electronics business I bought in Hobart together with New Zealander Bob Duncan, where I created the Tasman Turtle educational robot, was also a Dick Smith supplier. So, I sold Dick Smith electronics products instead of Radio Shack products. Unlike Radio Shack, Dick Smith sold hobby electronic kits for DIY enthusiasts to build their own esoteric gadgets like CB radios or metal detectors. Prior to that, hobbyists usually followed instructions from a hobby magazine, but had to go searching for all the components. The Dick Smith kits had all the parts and instructions in one convenient white cardboard box. Dick Smith became a highly visible and successful adventurer and entrepreneur in everything from making movies to flying solo helicopters around the world, taking off from, of all places, Fort Worth, Texas. These days Dick Smith is a generous philanthropist, activist and involved in federal politics.

Naturally enough, one venture we did in the early days to sell more Turtle robots was to supply it as a kit robot for hobbyists.

There were no robot products anywhere in the world at the time, and the Tasman Turtle electronics construction kit was published as a feature in one of the top Australian electronics hobby magazines called *Electronics Today International. ETI*, front page and all. Sales showed that it clearly filled a gap in the market. It generated interest greater than anticipated. Dick Smith had declined to take it on as one of his products when I met with him in Sydney one time.

One day in 1983 we received a confident-sounding call from an Uwe Mèffert, a Swiss entrepreneur based in Hong Kong, but the call was from Brisbane, Queensland. He had a copy of *ETI* and was reading the robot kit article. His interest was immediately piqued, such a novel electronics product was the Turtle. He wanted me to build a robot for him. He had no hesitation to say that if I did not he would copy the Turtle anyway!

Arrangements were made for me to fly to HK. The only international travel I had done until then was to New Zealand, but I had my passport and within days I had booked a flight, arriving at night in the amazing old Kai Tak airport, a colonial relic. Flying in lower than the high-rise buildings lining the approach like sentries guarding the runway, peeking out of the plane window into private apartments like a floating voyeur. To see the giant clockface across the way, smaller than the 60-foot Duquesne Brewery Clock I was to see later in Pittsburgh, the world's largest, but big enough. And surprised by the dazzling but unblinking neon lights. No flashing on and off so as to not interfere with air traffic control light signals. Then hit by the heat and humidity once outside, in a poorly lit, littered, exhaust gas smelly, arrivals area of people and taxis that was as accessible and convenient as no modern airport is. I have flown into Hong Kong's new airport since it opened and cannot remember a single memorable thing about it.

Tall, aristocratic in a European way, Uwe was a remarkable inventor cum entrepreneur. Like everyone I was enthralled with the then popular three-tiered Rubik's Cube. Uwe unbelievably had invented four and five-level versions and dozens of other games and puzzles. Back to basics, like I did always as a child, I took the toy apart, broke open one of the cubes; I'm still astonished by the complex inner 'wheels within wheels' mechanism straight off Ezekiel's chariot. What Uwe wanted, what everyone wanted as I was discovering, was a hobby robot. I was with Mim by then and Mim was soon to join me as she did on all of the early international adventures. So, we built the robot that was to be called Elami. It had a four foot tall, bulbous, attractive robot body, cute, curvaceous and colourful as Asians like their toys and dolls. On its semicircular head, vestigial of the Tasman Turtles, was a small computer screen featuring a talking face synchronised with the speech synthesis. It was designed and constructed in a messy, unair-conditioned sweat shop at the top of a plastics moulding factory in Aberdeen, outside of Hong Kong proper.

Uwe was connected to key people and suppliers in Hong Kong. We incorporated prototype components just emerging into the market, like 3.25 centimetre floppy disc drives. I was introduced to an electronics manufacturer called Keystone Electronics owned by Dr Lui, who fabricated products for Radio Shack among other things. I learned a lot about the importance of people networks and connections at that time. One of the most creative toy manufacturers in the world was Tomy Corporation in Japan, with some of the most innovative toys I have ever encountered. My favourites were a two-wheeled motorcycle called Air Jammer that had a piston engine powered by compressed air with a pump like a small bicycle pump. Another was a large wheel driven internally, so that it took off by itself under its own steam. Then there was Armatron, a toy pick-and-place robot like the ones used to manufacture cars. An executive at Tomy once told me that sales of these amazing toys were only mediocre because few customers appreciated the cleverness in their toys. He said they were examples of toys that adults like me buy.

Elami was prototyped for Tomy and demonstrated at the Tokyo Toy Fair in 1983. Beautiful Japanese women, students earning extra money, were hired and scripted to spruik it from a stage. Such was the fascination with this never-before-seen machine that industrial spies were everywhere trying to photograph the innards each time the back was taken off to replace the large gel cell batteries. I secretly took photographs of these photographers secretly taking photos. Elami finally reached the market in the different form of Omnibot, not one robot but a series of increasingly more complex designs, but still with the ever-present

Tasman Turtle-like transparent, hemispherical dome head, a technology atavism. A reduced version of the original robot was also produced with the name of Elami Jr by Robotland in Hong Kong.

During this time when we were presenting Elami in Tokyo, at an assembly with Mitsubishi, another of Uwe's contacts, around a huge boardroom table in a meeting room in Japan, I was invited to demonstrate the Tasman Turtle. About 20 besuited executives plus a camera operator at the far end of the table watched my robot centre stage performing on the boardroom table surface. Speaking in English and interacting with everyone I put the robot through its paces. Just for this demonstration I had programmed the robot to respond to Japanese instructions. At one stage I gave the microphone to a nearby person and asked them to speak to the robot, to tell the robot to move around using Japanese words for left, right, forward and so on, which he did. The room went quiet. They asked me to wait, which took a never-ending ten minutes. Eventually an elderly man in an immaculate silk business suit entered the room, clearly peeved at something, and they asked me and the Japanese speaker to do it again. The man said nothing then left just as abruptly and I continued with my presentation. Afterwards I asked what the camera was all about. I was told that no one spoke English in the room, despite everyone nodding on cue and chuckling when I had clearly made a joke. After I left, they planned to replay the video while someone translated everything. I asked who the gentleman was who came into the room. They said it was the chairman of Mitsubishi.

Dr Lui was interested in robots too. His connections with Radio Shack led to my first introduction to Jack Tramiel, head and founder of Commodore Computers, the world's most successful PC company at the time. That morning, I arrived at Keystone Electronics on Hing Yip Street, which appropriately means Big Business in Cantonese, and all the factory girls stared and giggled at me as I walked through the production line to meet Dr Lui. I asked Dr Lui if they had never seen a Westerner before, and he said no they were giggling at my European features – they had never seen such a big nose. He then arranged to take me to meet Jack.

A different time I was staying at a cheap hotel in Hong Kong and became seriously ill. Believing I was having a worse-than-usual asthma attack I had bought normally prescription medication over the counter for self-treatment, but it worsened. I remember hanging over the air-conditioner in the room to breathe fresh, cool air. Eventually I called Dr Lui, who called an ambulance and fathered me through several days of treatment in hospital for what was probably Legionnaire's disease caused by legionella in the air conditioner. Dr Lui was a kind and caring man. The hotel was a far cry from other places I stayed at over the years: the popular Peninsula Hotel with its legendary brunches, or the colonial era Victorian Barracks with its huge airy rooms and high ceilings, built when space was not at a premium. The next day Mim called the hotel to talk with me.

'Can I speak with Allan Branch please, a guest there?'

'He not here. He in hospital.'

'Oh! What happened?

'He sick.'

'Oh dear! Is he okay?'

Incredulously, 'He in hospital!' Like what did she not understand?

I was in hospital for several days, finally released all clear. Dr Lui visited regularly, unbelievable kindness, and the treatment was as effective and professional as any Western hospital. No costs.

Shortly after meeting Jack, I received a call from him in my hotel room offering to give me a job in Dallas at their research centre. He disclosed that his company Commodore Computers had a secret robot project that had come unstuck, and he wanted me to take it over.

The only robots ever that I was jealous of, that I did not develop, were the walking robots of Marc Raibert. I'm still jealous. Now and then I contacted him, suggesting unsuccessfully each time that he let me work with him. Bald, casual, bespectacled, never without a brilliant smile, dressed always as if he is about to fly to Hawaii, happy with his life's work. Marc was at CMU when I was visiting there and I saw his first attempts and eventually astonishing successes at a one-legged hopping robot, then a four-legged horse-like galloping robot. His work is as fundamental, critical and pivotal to robotics and artificial intelligence as anything that has been done, in my opinion. From CMU he moved to MIT in 1986, and his dynamic walking robots progressed to two-legged versions. Balancing and walking on two legs is entirely more difficult than on four legs. Or six or eight or a hundred. He discovered and implemented the surprisingly simple but elegant feedback equations that allow a thing to balance and walk, on one or four or two legs, even when disturbed by pushing or bumping into it for instance, and eventually running and jumping. Terrifyingly animalistic. Not helped with names like Big Dog and excessive pneumatic and mechanical noises. Not surprising that the Rambo-types at DARPA picked up on his work to contract defence versions. He moved from academia to private enterprise in 1993, incorporating Boston Dynamics, which in turn was acquired by Google in 2013, then to Softbank in 2017, then by Hyundai Motor Group in 2020.

One reason, maybe the main reason, maybe even the only reason I quit robotics in 1997 was because in my opinion there was no market for them. My goals were commercial. There were a few niche markets – the ones I had dominated were the toy and hobby robot sectors, others were the education and research markets – but otherwise no one needed or bought robots. A few small markets for disaster areas or bomb disposal maybe. A robot to me is an artificial person and extremely complex and difficult, beyond our skill so far except in limited cases. This was a tenet that Google may have discovered.

At about the same time as Marc was having his fun, and it was real fun according to him, I had my fun creating Mr

Walker, named after the lead character in my favourite comic books growing up, The Phantom, Ghost Who Walks. It was a dynamically balanced two-legged walking robot, and of all the hobby robots I have developed it is my favourite. Marc's robots cost tens of thousands of dollars to buy. Millions to develop, Mr Walker costs a few tens to buy and was developed over a weekend. If you pushed Marc's robots, they righted themselves. If you pushed mine, they fell over. Much more simplistic but no less animalistic in looks and action than Marc's walkers, Mr Walker moved its centre of gravity across a frame in the torso to locate its centre of mass alternatively over the feet as it walked by lifting the opposite leg. It could walk up steps and all that cool stuff, and had a few of the other applications like speech from my previous robots added in. The head was still a half-circle like the Turtles but no longer transparent.

The other connection I made through Uve Mèffert and Dr Lui was Bernie Appel. Given the market depth of Radio Shack I wanted desperately to have them sell one of my robots. Back at our head office in Tasmania we all felt that Mr Walker was the robot to present to Radio Shack.

The Radio Shack product selection process in Hong Kong was coordinated by their Japanese purchasing affiliate A&A International and always attended to by Bernie. Bernie's onefold skill was making instant and accurate commercial decisions about what would sell.

Getting a few minutes to present a product to this giant international electronics merchant was competitive and a special event. My contacts and track record of robot products throughout Asia had secured just such a spot. Mr Walker was still a raw prototype, though, with visible machinery and circuitry like a skeleton from an anatomy laboratory. Our technical people loved it, but merchants and salespeople have no vision. No good showing them something incomplete and describing what it will become. We have to turn it into a finished product. My friends Alex and Ron were skilled at making fibreglass kayaks. We quickly built a fibreglass body for Mr Walker and prepared to travel with it to Hong Kong. But there is another participant in this story.

Australia is sports mad. Our local sport is called Australian Rules Football, or Aussie Rules for short, probably a form of ancient Irish football. It's a rough, dangerous, rapid, contact game that involves kicking the leather, oval-shaped ball between tall posts at the ends of the oval-shaped field called, wait for it, an oval. The field is a cricket ground and the game evolved as a sport for cricketers to keep fit through the winters, which is off-season for cricket. The name is a misnomer. It feels like there are no rules and no player wears protective gear. Aussie Rules sportspersons are our heroes, up there in the stratosphere, untouchable, so it was a shock to receive a call from Ron Barassi one morning.

Ron was a legend, the first footballer to be inducted into the Australian Football Hall of Fame, medium height, handsome, charismatic and famous for transforming Aussie Rules into a modern technical game. Ron passed away as I was writing this, so sadly I had to go back and edit the present tenses to past tenses. Like many sports celebrities, after retiring he used his fame and name recognition in business. He had joined real estate developers and investment entrepreneurs Harris Brothers, and was searching for promising startup enterprises for them. We had caught his attention. Ron came to visit us and immediately they invested much needed cash in our company. They heard of the impending Radio Shack opportunity and wanted in. But they insisted that their representative had to fly with us and participate in the product presentation in Hong Kong. No problem. Having one of the mature, professional Harris brothers would add commercial expertise to our troupe.

News of this was in the air, as it was such a big deal for a local tech startup. Artistic designers at the Federal Department of Science and Technology offered to spruce up the plain white robot body, adding some flash and flare, an offer that was readily accepted. It was a last-minute offer, so through the afternoon and into the early evening before the early morning flight out of Hobart, they took the robot to their studio in the city and made it beautiful.

While we were waiting for it to come back to us, a call came from the government officer saying the robot had gone missing. We were devastated, he was devastated, we were all distressed. I called everyone back to the workshop and we decided to work through the night to try to make a second Mr Walker from spares and bits and pieces lying around from the prototyping process. Make a robot that took six months in six hours.

Apparently, the government officer had finished his design work on the robot body, then carried it down to his car in the open car park under the building, but forgot his keys. He hid the robot in a corner and raced back to his office to get his keys and when he returned the robot had gone. Police were called, news flashes were broadcast non-stop on radio and TV, and a reward was offered. Around midnight, a call came to the police by a teenager who had the robot, offering to return it but wanting the reward. His parents were not impressed when suddenly flashing police cars stormed their house. We got our robot back. The boy was the one who stole it, and there was a coin-sized slot cut into the head where the teenager had decided to turn this irreplaceable prototype into a money box. With the robot safely back, and everyone relieved and heading back home to bed, we then packed and prepared to proceed to Hong Kong bleary-eyed early next morning, to turn the robot into a money box in a different way. The dedication, loyalty and motivation of the team in this time of crisis is something I have never forgotten. I think the government official has never forgotten the night either.

But the Harris brother who was to travel with us was not at the airport next morning. It is a small, one level airport and we searched everywhere, concerned again. It was becoming a litany of errors. He realised at the last minute that he did not have a passport. So his brother Ken, who had a passport but had not spent any time with us and knew nothing about our company, products, technology, connections, was travelling instead. Ken was taking the next available flight and would connect with us over there.

At the presentation in Hong Kong, Bernie and his colleagues were visibly impressed with Mr Walker. There was nothing like it on the market, and it was rare to see their enthusiasm so evident. No poker playing. They went over the pricing several times to determine what it could be mass manufactured for and sell for. Before the meeting Ken instructed us that he would control the contractual terms and pricing negotiations. He stuck to a price that they could not do, and try as they might Ken was unrelenting and the deal fell through.

Mr Walker became a prized demonstration, often set walking around the stage at international conferences and workshops, but despite being my favourite robot, never made it to market.

In 2014, a couple years before he died, I caught up with Bernie Appel to have lunch with him at his private club in Fort Worth. We reminisced over the Mr Walker episode and we took a picture of both of us, global technologists, proudly displaying our simple minimalistic flip-top cell phones. He still maintained his cherished office in the Tandy Twin Towers in downtown Fort Worth with mementoes of his life and career, and seemed genuinely excited to have a visitor.

CHAPTER 32

FLYING HIGH

My first international flight was to New Zealand from Hobart, to escape my feelings for a woman I had an unhealthy obsession with. That we had for each other. Once there I discovered New Zealanders often flew to America and I envied that. It was at a time when international travel was still uncommon, and if done at all was frequently by ship. My friends in Australia who had been lucky enough to travel had all gone to London, the traditional coming-of-age adventure of university graduates, honeymooners and wealthy young adults before they returned and settled into their banal lives, rarely to travel again. It was Australia's Grand Tour. Travel to the New World seemed more interesting to me than the Old World, and I envied those New Zealanders.

Not surprisingly, the next time I flew internationally was to America, not trying to escape anyone or anything, but chasing the dream that was to be my robotics career. It was on New Year's Day 1981 and as I flew over the International Date Line I celebrated New Year's Day a second time, before arriving in scintillating San Francisco, warm even in winter. Since then, I have circled the globe at least once a year, sometimes two or three times a year, at times holding simultaneous two or three round world tickets to pick and choose individual legs as my plans and itinerary altered according to the opportunities generated along the way. I was a nightmare client for my travel agent Jane Johnston, or Jane Evens as she was then; her company also changed names and is now called Travel With a Cause. It was as if travelling was my cause because Jane and her company feature heavily in this chapter.

From San Francisco – after doing the obligatory Pier 31 and Embarcadero, little cable cars and windy Lombard Street; discovering innovations not known in Australia like four-way stops at traffic intersection; relating to the little boxes on the hillside of South San Francisco like in the song; and touring a population the size of Australia's just in that one Bay Area – I flew on to Boston. Boston was a different world from California, more like the Old World, and I could have been in London after all; no Spanish, no warmth, no massive multilane highways. Irish, icy cold, narrow busy streets, crowded buildings. The hotel room was uncomfortably hot to compensate for outside's freeze. Yet such brilliant blue sky, like the ultramarine of grandmother's laundry blue; in Australia with a sky that blue it would be as hot as the hotel room. Instead, the cold snaps your face and stings extremities. Then to Quebec, Canada, and colder still if that were possible, where I could have been in France. All of it remnants of European colonialism.

My friend Ron Davies, a student friend at university who ended up his whole working life with airlines, initially with TAA, Australia's government led airline back when we had a two domestic airline policy; the other was a private enterprise called Ansett. The policy was established to break a monopoly held by ANA before that time. Curiously ANA started as a small local Tasmanian airline, but by the 1940s had grown through amalgamation to dominate Australia's air travel and freight, with the last transaction being to sell to Ansett. In time TAA changed its name to Australian Airlines, but then was bought out by Qantas, and eventually was completely integrated into the international carrier as Australia relaxed its policies and deregulated the industry. In turn, deregulation led to the demise of Ansett and the admission of anyone and everyone into the Australian market. Ron was there through much of it.

When I mentioned to Ron that I was petrified of flying, had in fact even avoided a flight now and then because of a premonition of disaster, luckily never coming true, he arranged for me to fly from Hobart to Melbourne in the cockpit with the pilots, or as he called it by the more fortunate name, the flightdeck. As I flew more and more my fear increased rather than waned. Riding that day up front in the little pulldown 'jump seat' behind the cockpit door was an experience. Seeing the take-off and approach, seeing for the first time ahead in the direction we were flying instead of out the side looking at where we'd been, the controls with landing lights and systems, the competence and calmness of the pilots, who explained things to me as if I were a student pilot, built my confidence in and cured my fear of flying. I even did a few hours training for my private licence in a Cessna 172.

My friend Herman Stevens in Holland, an engineer with the Royal Dutch Airlines KLM, took me and my sister Dianna through a 747 Jumbo that was being maintained and refurbished at Schiphol Airport outside of Amsterdam. Inside the stripped plane, new fuselage ribs were being installed to strengthen the body, every second one this time, the alternative ones next time. He explained that after one team designs the aircraft and put it in production, another team does the design calculations all over again as a safety measure, and in that case they determined that another millimetre or two width of the rib sections would be safer. I discovered things I never knew, like the separate engine in the back called the auxiliary power unit to power the plane when the main wing engines were off, the Black Box, which is orange, and which manufacturer I met a few times in Anaheim, California negotiating a deal over their Functionoid robot. My sister and I sat in the pilot's seat taking pictures like prize peacocks.

On 15 January 2009, I was in New York, as usual, and was watching TV when live footage showed a pilot who'd navigated a plane that had lost all of its power from ingesting birds into its engines on take-off from LaGuardia airport, to a perfect landing in the Hudson River. The pilot was Chesley Burnett Sullenberger III, who luckily has the nickname Sully. The river was flat and calm, like a runway, not churning as it can be. He did it as if this was an everyday routine activity. I doubt I have ever been more proud of any individual. To my mind it was a feat up there with the safe return of Apollo 13 with its three crew after their spacecraft had a catastrophic accident and lost all power, or the safe rescue of the 13 soccer kids from a flooded cave with no way out in Northern Thailand. The unlucky number 13 features prominently. The return mission of Apollo 13 took four days; the return mission of the Thai students took 18 days. Unlike those rescues which had time and unlimited resources to plan the rescue, Sully's flight took one and a half minutes and he had seconds to decide how to save his 155 people on board. I watched this live on TV from a few blocks away in New York with the Hudson River in view.

Gordon Bellamy was a dear friend, as eccentric as they come, and a talented artist. He was an animation artist for the likes of Disney, Hanna Barbera, MTV and others. I met him in Dallas when I took on the job with Commodore Business Machines. Commodore had hired him to design the robot bodies in the project that had floundered and which I was rescuing. That would have been a dream position for him. In the little studio he created in a back room I discovered two full-sized foam models he had fashioned for them. One was a traditional robot design, a cylindrical rubbish bin with a head and arms, eventually adopted as Chester; the other was a freakish design, a cross between every sci-fi alien ever seen. He was having fun at Commodore's expense. I ignored that one and went conservative. It was interesting for me that his two robot models encapsulated my twin values, science and technology on one side, art and nature on the other. Gordon and his partner Dabni became great friends, and I often stayed with them in California when they moved there, and in New York when they moved there.

After leaving the Commodore Gordon and I travelled across country to California for our Jack Kerouac road trip. He wanted to visit his children and ex-wife in San Francisco, then do reconnaissance to find work with animation companies in Los Angeles. After travelling north from Dallas to Interstate 40, we crossed the western Texas panhandle, with country soil as yellow as sunflowers, to namesake Amarillo, which means 'yellow' in Spanish, tumbleweeds like giant hairballs or oversized thistle seeds or a ball of string that you cannot unwind, across foothills and canyons and mountains, past stunted, disfigured mesquite trees to the border into California. Then diverting towards Bakersfield, home of the Bakersfield country music scene of Wyn Stewart, Buck Owens, Merle Haggard and Bonnie Owens, with their distinctive syncopated backbeats, we took other roads trying to connect with Interstate 5. We were travelling according to Gordon's memory, taking back roads, searching for the interstate to take us north-west in the direction of San Francisco. He was relating his previous trip when one time he got lost and came to a small town actually called Lost Hills. Gordon, who grew up in Corpus Christi at the coast of the Gulf of Mexico in Texas, loved Mexican food, or more accurately Tex-Mex food. He said that at Lost Hills he had genuine homemade Mexican food, with tortillas made from scratch, fresh farm ingredients, home-ground cornmeal – just the best Mexican food he had ever eaten. Unbelievably, as he was saying this, we saw a sign for Lost Hills on the side of the dirt road we were on, and we drove into town. Sure enough the same family restaurant was there near the entrance to the town, which could have been from a Western movie. They made homemade tortillas and tacos for us, soft yellow corn, shredded beef, sweet hot salsa, all of which were the best I have ever eaten too.

We had already passed through Bakersfield, a city 100 or so miles over the mountains if you drive north out of Los Angeles and down the steep Grapevine, made famous in the Charlie Ryan song 'Hot Rod Lincoln'. The city was the scene of a breakaway country music scene called the Bakersfield Sound in the 1950s, with Buck Owens the relocated Texas singer considered a local, becoming one of the greatest international stars of all time. It's called The Grapevine because of the grapes, but it is winding and deadly like twisting ivy vines. Buck's unique style has been copied by many, not the least of which were the equally prominent Dwight Yoakam and Merle Haggard. Merle later married Buck Owens' ex-wife Bonnie Campbell Owens, who was also a prominent country music artist. Not much later I actually spent an evening backstage with Bonnie while nervously waiting to do a radio interview with Merle after his concert at Tramps in New York. She talked about her continued touring with Merle, and her wonderful new husband, a businessman, back home. The support band was Sleepy LaBeef, who I'd never heard of. A couple of his band members were there chuckling at my flirting with Bonnie. Sleepy, who was definitely beefy, but then became slim and healthier looking, was a loud, heavy-duty, rockabilly artist from Arkansas, whose real name was LaBeff. There are later images of him, bespeckled, jowly and kind-looking, where he could have been a physician. Afterwards I went out and bought a Sleepy LaBeef album and played it once. As a Texan, Gordon was as familiar with country music as I was and equally enjoyed this side to our trip.

Another time in New York I went to meet Virginia to go out for dinner. She was a secretary at Atlantic Records' head office and while waiting for her to finish work I heard a great song playing in the background. I met her boss Ahmet Ertegun, the co-founder of Atlantic Records, and at one point did a small project for him on desalination to make fresh water. He was originally from Turkey and anything to secure water supplies for his home country was dear to his heart. When I left with Virginia that day, someone raced out and handed me a copy of the unreleased new CD with the song I had commented on, 'The Words Inside', by their newly signed band Boy on a Dolphin.

Back on Interstate 5, Gordon and I continued and eventually arrived in San Francisco, my second time, back and forth across the bay, past famous gaols redolent with the strains of Johnny Cash and escape tales of Birdmen, and horrors of capital punishment, north to the Russian River, where he reconciled with his ex-wife, met her new husband, and reunited with his estranged teenager kids, a boy and girl.

While in Munich in 2018 I received a teary call from Jane Johnston, my travel agent of 30 years and now owner of Travel With a Cause in Hobart. Jane knew of my company turnaround activities and TWAC was in trouble, she said. The travel industry had changed dramatically over the years, and she was concerned about the business. Should she close it down, or could I help? For several years I had been on her board of directors, and I could not do more from that position than I had been doing.

'Hi, Allan, I'm in trouble. My company's in trouble. I work harder than ever and make less profit, and feel like I should close the business down, but I don't want to. It is my life. No one seems to need a travel agent anymore, because they can do it all online.'

'All businesses can be turned around, Jane. There are successful travel agencies around.'

'Allan, is there any way you could help me?'

'I can only do more if I work hands-on, in-house as the CEO, like I normally do for my turnarounds.'

'Great, could you do that?'

'Yes, but you have to understand what the CEO is. It has the word "chief" in there. It is the top management position, and it means that you would work for me, take instructions from me, even though it is your company.'

'I'm happy to do that.'

'Everyone says that, but in my experience owner-founders can rarely actually take instruction from an outsider, and it leads to problems. If we do this, you would have to sign an agreement to it.'

'Great, I can do that.'

And she did. I flew back to Hobart to undertake the transformation of TWAC, the only not-for-profit travel agency in Australia. TWAC is dedicated to donating 100 per cent of its profits to worthwhile causes in developing countries, such things as education, water supplies, health, infrastructure. Not-for-profit does not mean no profit, and charitable causes need money from somewhere, often grants, sometimes small fundraisers like auctions and flea markets, other times through traditional commercial activities. TWAC was a full service national and international travel agent, earning its revenues through commissions and sales of travel products.

Before I started, she said one thing more: 'I cannot pay you.'

'Oh!'

Winter was coming on in Munich, but I flew to Hobart which was enjoying late spring. I set up office at TWAC, rolled up my sleeves and got down to business. We separated divisions and tasks within the company, researched the industry and its changes since the internet, installed strong governance and HR practices, ramped up our social media marketing, and analysed our product and pricing offerings setting different sales priorities.

Within a year we had doubled profits, opened a second office and won an international award as the best travel agency in Hobart. Once done, I flew out, heading back to Germany, getting as far as Bangkok when COVID-19 hit.

On Australian government advice that if any expats could, they should return to Australia where it would be safest, I turned around and flew back to Hobart. The Australian government banned all national and international travel, so TWAC's revenues dropped to zero overnight. I rejoining TWAC. Still unpaid.

After a board meeting where the board resolved to try to keep operating, I called everyone to a meeting and said, 'Forget everything we did last year. The rules no longer apply, so we have to do a new transformation of TWAC.'

One thing we had done the year before was remove loss leading sales and through the work of marketing manager Amber, we added some local volunteer group experiences. They had only been offered now and then, but their existence was a lifeline. The decision was made to do these weekly, as they did not entail any air travel. Because of the travel bans, many international students and tourists were trapped in Australia. Congregating for concerts and dining and similar social events was also banned by state and federal governments. As alternative opportunities for entertainment dried up, they found participating in our small volunteer tours exciting. We revitalised revenues, less than before but with higher margins and doubled profits again. TWAC received the Best Travel Agency in Hobart again, an award won by TWAC for six consecutive years now, and in 2022 was awarded the most sustainable tourism operator in the world by peak industry body Skäl.

In the past I had worked with not-for-profit organisations, even a bit of pro bono, but this extended experience with TWAC brought me back home, to my beloved Tasmania, Van Diemen's Land, to more interaction with family and friends, and a change of direction to more not-for-profit engagements.

Most of my time between engagements is either in New York or Germany, where I begin my search for my next turnaround. This particular time I had been in Zurich from where, the day before, 10 September 2001, I had arrived back in New York. Still sleepy-eyed from jetlag, I was at Zack's place in the Bronx when we were all suddenly glued to the TV. Something between a nightmare and a horror movie was on the screen. The images of one of the twin World Trade Center Buildings in downtown Manhattan, two blocks from where I normally stayed with Gordon and Dabni, was on fire and smoking. As we were watching, unbelievably, incomprehensively, a plane flew into the second building. It seemed to go straight through. We realised then that the fire and billowing smoke of the first one had also been an earlier plane. Then, as if nothing could be worse, the buildings collapsed, slow motion at first then hidden by dust as thick as a sandstorm from pulverised concrete. People were in shock, crying, struggling to breathe, racing out of the dust storm as if they had been fire bombed in Vietnam or a Top Gun walking out of the smoke from a plane crash. The bright blue sky turned the colour of slate.

Recovering some composure, I called Gordon and Dabni, and heard a message on their answering machine that they were okay and had moved out of the apartment. Debris had started falling on the roof of their apartment, a foot thick, and it was unsafe. If I had called a short time later, I would have heard nothing to calm my angst as the power and telephone systems in lower Manhattan collapsed.

Later, when life was starting to return to normal, Gordon was asked by his Animation Industry Body in Los Angeles if he could write a story for their magazine *The Pegboard*, on the day of 9/11 from the perspective of an artist member who was there as it happened. As fantastic as Gordon is at drawing, he is not as comfortable with words. Postcards from him usually have a picture instead of a written message. He drafted his story but asked me to edit and proof it for him. The article was published in the magazine in two monthly instalments, including harrowing sketches Gordon made on the day. The apartment is the top floor and has a skylight from which the World Trade Center Buildings can be seen, and we have photos of them on fire on that morning.

Life did not return to normal, though, not really. Every time I was on the streets of Manhattan, from where planes landing at any of the three local airports can be seen on their low approach, I ducked reflexively. It was six months before I flew again. I was scared, and I hate to admit that on that flight, in line at the departure lounge about to board the plane, I checked out every turban, kaftan and beard in the passenger line.

Gordon's story, edited by me to be in his voice, was published in abridged form in two parts in *The Pegboard*, in March and April 2002.

What I consider more than anything about flying high all these years is that it was always alone. While I was living that life of non-stop adventure, gallivanting around the world, I was a lonely man. Always leaving somewhere, someone, going to somewhere, someone. But the ride was solitary and introspective as I wondered constantly where I was really heading in life. I was tempted to name this book *Flying Solo*, such was the gravity of that feeling. It was never boring, though. The first time I crossed the Pacific I stared at the ocean from 40,000 feet, imagining one of the early explorers or modern solo sailors down there, facing the elements on the currents like the Gulf Stream as we were facing the slip stream in the firmaments. On two occasions I sat next to a smart, beautiful single woman, enamoured with each other for the 14-hour flight, under the blankets, comparing notes, like one might if suddenly the world was about to end, safely never to see each other again. Flying high. A mile high.

CHAPTER 33

THE RUB

There are no guarantees in life. No fairness, justice, karma, no atonement. Life should come with waivers, like the disclaimers and risk assessments in contracts and prospectuses. Insurance companies do that – try their best to eliminate anything that might actually happen to you. 'Why did this happen to me?' is the wrong question. There is no 'why'. It's just the universe doing its thing.

So, I do not ask, 'Why was I so lucky to have had the life I had?' or 'Why did those things happen to me?' None of it meant anything to the universe. It will disappear when I disappear. I have no religion. I believe there is no reason to it all. I believe that when you die you die. But in a solipsistic sense, I do ask if any of it meant anything to me. I get three discoveries and one answer. This was probably the unconscious motive for writing this book.

For one, I know what a robot is and what it is not. Everyone who has delved into robotics has their pet definition, some of them tautological or boring. Formal definitions often contradict. *Britannica*, for instance, says, 'any automatically operated machine that replaces human effort, though it may not resemble human beings in appearance or perform functions in a humanlike manner', while Merriam-Webster says almost the opposite: 'a machine that resembles a living creature in being capable of moving independently (as by walking or rolling on wheels) and performing complex actions (such as grasping and moving objects)'. Nothing about intelligence or reasoning or decision making like we prefer to imagine robots, smart machines.

My favourite definition is Joe Engelberger's. 'I cannot define a robot, but I know one when I see one.' That is a waiver if ever I saw one; but he should know, he invented the world's first robot. Mine is different. I don't hedge my bets. 'A robot is an attempt by humans to duplicate the behaviour of intelligent organisms ...' Here an intelligent organism needs to be defined, so I continue and default to Rodney Brooks. 'A robot is an attempt by humans to duplicate the behaviour of an intelligent organism, an intelligent organism being by necessity one that displays mobility.' That excludes plants and things moving solely in wind or water currents; for instance, bacteria or spores. It includes butterflies, and worms, and mice, and men. So, it does not include talking elevators, guided missiles, talking AI systems like the current chatbots and their large language models (LLMs), or remote-controlled toy robots. It includes driverless cars, automatic vacuum cleaners and Big Dog. Maybe it includes Engelberger's robotic arms.

Working in robotics, designing machines to function like living biological organisms, required me to call on everything I had learned from medical school, which was not enough. I studied psychology, neurology, philosophy, natural intelligence, group and mob behaviour, language *à* la Chomsky, senses other than those in humans, vision systems, hearing systems, proprioception. This clashed with most of my colleagues and competitors who were mathematics-oriented, computer scientists and physicists with little to no biological training; or interest. They preferred, relished even, complex approaches. Brute force. I preferred taking advice from nature with its elegant solutions. In the end, building fake humans possibly also helped me to understand real humans better.

For a second thing, I know what a company is and what it is not. More accurately, I know what a company should be. For this I had to go back to the beginning of companies, to medieval Venice, tracking their evolution through Holland and England, to modern corporate laws and structures. Like democracy and justice, companies are a human invention, a concept that does not exist in the universe otherwise. Unless there are Martian corporations. As a concept, it is like being a god, because creating a company is creating an entity out of nothing, a new individual, non-living, of course, but a thing like a person, nevertheless.

In fact, a company registered in a different state in the USA is sometimes called an alien.

My discoveries were again clashed with the established wisdom, and definitions at times were again contradictory. Wikipedia says, 'A company … is a legal entity representing an association of legal people … natural or judiciary … [who] share a common purpose to unite to achieve specific, declared goals.' The definition has an interesting, subtle tautology: the rose is a rose kind, when they say it is a legal entity that can be judiciary, which implies a legal entity. Investopedia has an even weaker definition and a more glaring tautology: 'A company is a legal entity formed by a group of individuals to engage in and operate a business.' Sounds to me like they are saying a business is a business. My definition is that a company is an artificial person whose purpose is to sell stuff. Most dictionary definitions now describe a company that same way, but they never used to. An example is Cambridge, which says, 'A company is an organisation that sells goods or services in order to make money.' It must be axiomatic, though, because there is that 'a company' is 'an organisation' again. And nothing about shareholders or what their shares really are.

Unlike a real person, a company is not a conscious being in control of itself, so companies need a head. The 'head' is a board of directors, whose function is to ensure the company performs in the best interest of its shareholders, whose interest is to protect and preserve their investment and to grow its value. Companies have their own set of laws, and a focal point of those laws is around how they are led. A director of a company therefore has a tough job, required by law and regulators to act outside of themselves and their personal interests. A direct and unfortunate consequence is that shareholders of large listed companies today are as often as not other large listed companies. So, shareholders are often faceless and interested only in profits. Directors and executives, hired to look after the best interests of these types of shareholders, therefore, are profit driven and must be by law. Those focussed on nothing else rise to the top in this structure. It has been discovered that an inordinate percentage of corporate leaders are sociopaths, immune to emotions and feelings about the consequences of their behaviour. I can think of the tobacco industry for instance.

A company cannot vote. For me, this is the most important difference. They therefore have no political power, nor should they. In my opinion, companies should not lobby or influence politicians or voters. Contribute to policy development, of course. Companies should not entice its employers to vote to their preferences. Companies should persist and be supported by being good community citizens, like any other person.

Like doctors who can fix bodies because they understand how they are constructed, knowing intimately, definitively, what a company is, how they function, provided me with the skill to fix them. Half of that knowledge came from building and running my own company, then the other half came from discovering the problems and causes of problems in other people's companies.

I am one of the few corporate executives who believe they should not own shares in the companies they manage. The exceptions, of course, are owner-managers of private companies. Anyone managing a public company, which by law must ensure a true and equitable market, meaning everyone investing in it has access to the same information as anyone else, all the time and at the same time, always has inside information. If they do not, then they are not doing their job. It means that they always have an advantage over external investors. It also means that they can influence share prices; for instance, by buying into a struggling company to falsely suggest to the market that it is worth investing in.

There is another reason markets are not rational. There are two types of shareholders. Those who build companies, backing good management, patiently wait for the company to grow and prosper and for the share price to follow suit, rationally. Then there are those shareholders who simply play the market to make quick money. Often called day traders. Their mindset usually affects share price independently of what the company is doing, like the Heisenberg uncertainty principle, or more accurately the 'observer effect'. Those companies are built on manipulation, not substance. Deliberately creating a false market is illegal, but is impossible to prove. While the manipulators seem to make money whether the share price goes down or up, the other stakeholders affected by their behaviour rarely profit.

A lifetime resulting in two discoveries is not much to show. But there is a third discovery. I know what I am and what I am not. At Zack's I was a frequent house guest. Downstairs in the house he designed and built in the Bronx, I had slept in the homemade single beds in each of the bedrooms. One room had been converted to an office, and the other two were still used for guests now that the kids were adults away from home. Like a lot of architects, one of Zack's hobbies was woodworking, and he constructed homemade wooden beds with neat pull-out trundles underneath for extra guests, and sleepovers when his kids were young. Despite it being the heart of greater New York, I recall one day lying late in bed and listening to the sounds of suburbia, kids playing, a ball being kicked, lawn mowers mowing, someone washing their car, doors slamming, a kid crying, neighbours chatting over the fence, others calling for someone, music flowing out of windows, trees rustling in the wind, bees buzzing around flowers, dogs barking at other dogs and birds tweeting to other birds, trains and traffic in the distance, planes and helicopters overhead, and I thought, 'The world is okay.'

Zack's middle child Paul had schizophrenia, what used to be called a split personality, but which can be a severe psychosis like hearing voices, which is what Paul said he experienced. He was capable of independent living with support, so worked in a regular day job, had his one-bedroom flat on the Lower East Side of Manhattan, and a girlfriend, eventually a wife. He received a weekly stipend from his trust fund, which Zack managed for him. Every Sunday, Paul would catch the subway to his parents' house in Riverdale in the Bronx, stay for the day and have lunch, and Zack would take him home in his car at the end of the day. Zack gave him his cheque at the end of the trip. I observed this frequently, sometimes going for a ride with them. This day, things were different; Paul and I spent some time downstairs in my room where I showed him what Zack and I had been doing in his aged care consulting activities. It was getting late and Zack was tired. He suggested Paul might like to stay over, to save him from driving all the way into the city late at night. Paul said no, and I could see that Zack was disappointed. I said to Paul that I could move out of my room into the end bedroom and he could use my room downstairs if that helped. Paul said yes, and he stayed over, much to Zack's relief. Next day, after he had driven Paul home, Zack came up to me with a mixed expression of wonder and amazement and asked what had triggered the suggestion the previous night. I said that I had slept in each of the homemade beds, which were all different sizes, and the only one left for Paul to sleep in was shorter than the others. Paul is over 6 feet tall, and the bed would have been uncomfortable, but he did not have the communication skills to say so. My bed was the longest in the house.

Several people have said I am a good teacher. I love teaching and appreciate the compliments. But teaching is restricted to the known. That is perhaps why being an autodidact is at times more fruitful than being a formally educated person. An autodidact has no knowledge of the boundaries of what is so far known in a field, so ventures further. I delve into the unknown, into unsolved problems. What is a robot, what is a company, why does Paul not want to stay over to please his dad? So, what am I? A problem solver, seeing causes and effects, trends, insights, consequences that others miss. The constant reward in my life has been the warm buzz from solving problems.

That is the third realisation I had while creating this book and being forced to think about what it all meant.

Most authors search for a cute quote to encapsulate the meaning behind their work or a chapter in their work and I kept wanting to use Shakespeare's 'To be or not to be, that is the question.' I know it is the most oft used quote in the universe, but that is because it has something about it. Life or death, basically. But my story is not about the choice between life or death, so I could not use it.

My story instead is about progress, not always in the direction intended.

In 1848 Phineas Gage was working on a railway line when he had a crowbar go completely through his skull, brain and jaw from an explosion accident. It damaged his left frontal lobe, which is a key part of the personality structure of the brain. It deals with planning, attention, reward and motivation. It is a part of the brain that contributes heavily to making you who you are. Amazingly, he survived, and recovered his intelligence and strength, but not all of his social skills. He appeared initially to be unaffected, but soon it emerged that his behaviour was such that he often did things against his own best interests, as if he had no control over tact and discretion, unable to assess the immediate risk from reactionary behaviour. He had an inability to behave in a way that would bring later and greater rewards instead of instant gratification. It makes me wonder what has happened to the frontal lobes of gamblers, or those with anger problems, or who smoke cigarettes. It makes me wonder what has happened with those shareholders who trade or vote irrationally, particularly when I compare the illogical behaviour and avarice of some shareholders at Krucible which resulted in their losing their investments with the short-term sacrifice made by my geology staff that time when they voluntarily became acting investment and fundraising staff way out of their comfort zone in order to secure a much greater later reward, saving the company and their jobs. Or the staff who volunteered to work overnight to try to fabricate a second Mr Walker when the original was stolen. It is that unselfish behaviour that convinces me to continue saving companies, even when the unworthy sometimes benefit too. Even when I sometimes get slapped in the face.

In Hamlet's unforgettable soliloquy, in the tenth line, after saying that all problems would disappear if one just died, he then says that as no one knows what happens after death, it may not be the end of troubles. Hell, dreams, nightmares, fire, damnation. 'To sleep, perchance to dream – ay, there's the rub.' It is not about living or dying. Not a question of 'To be or not to be.' Instead, it is about living large, as Moss Webster said in Michael Robotham's

Life or Death: 'One crowded hour being worth more than a lifetime of the commonplace or the mundane.' No doubt Robotham had read similar words in Tim Bowden's heartbreaking book about the shocking story of Neil Davis, extraordinary cameraman, who was from the town of Sorell, next to my home in Tasmania for 30 years, and who filmed his own death.

So, for me it is definitely about living, living large. But here's the rub: things always have a habit of getting in the way. Condensing all of this, I have learned from the misfortunes of the lives of my family and friends and colleagues and associates, and using deeper problem solving than they could ever have imagined or mustered, avoided their fates.

There are no guarantees in life. No fairness, justice, karma, no atonement ... 'Why did this happen to me?' is the wrong question. There is no 'why'. It's just the universe doing its thing.

ABOUT THE AUTHOR

Allan Charles Branch was born in Tasmania and attended multiple schools across Hobart, moving frequently as his family renovated homes. Though he originally aspired to be an artist, Allan's early jobs revealed a natural aptitude for science and technology. After finishing at Elizabeth College, he briefly studied medicine at the University of Tasmania before leaving to launch a high-tech career in robotics. Starting in 1979, Allan quickly gained an international reputation, developing smart mobile robots for clients such as Commodore, Radio Shack, and Moulinex across Asia, the USA, and Europe.

In 1993, he took the helm of Denning Mobile Robotics, a struggling NASDAQ-listed competitor, which he successfully revitalised over four years with high-profile clients like General Electric and NASA. Following his work in robotics, Allan found his management skills transferable to other industries, establishing himself as a specialist in global corporate turnarounds—a career he continues to pursue, though he now devotes much of his time to writing. While Allan has been published in both academic and popular media, this is his first book. He currently divides his time between Tasmania, the USA, Thailand, and Germany.

For more great titles visit

www.bigskypublishing.com.au